Weekend Projects for the Radio Amateur

Edited by

**George Brown,
PhD, MW5ACN**

Radio Society of Great Britain

Published by the
Radio Society of Great Britain,
Lambda House, Cranborne Road, Potters Bar, Herts. EN6 3JE, UK

First Published 2008

© Radio Society of Great Britain , 2008. All rights reserved. No part of this publication may be reproduced, stored in a retrieval system, or transmitted, in any form or by any means, electronic, mechanical, photocopying, recording or otherwise, without the prior written permission of the Radio Society of Great Britain.

ISBN 9781-9050-8641-2

Cover design: Dorotea Vizer, M3VZR
Subediting & Typography: Chris Danby, G0DWV.
Production: Mark Allgar, M1MPA

Printed in Great Britain by Latimer Trend of Plymouth.

Publisher's note
The opinions expressed in this book are those of the authors and not necessarily those of the RSGB. While the information presented is believed to be correct, the authors, the publisher and their agents cannot accept responsibility for the consequences arising for any inaccuracies or omissions.

CONTENTS

Preface vii

Aerials

Two-element, 6m Quad aerial 2
A portable three-element 6m Yagi aerial 4
Simple aerials 10
Aerial maintenance 14
A J-Pole aerial for 50MHz 17
The G3HBN portable magnetic loop 20
A UHF corner reflector aerial 25
A tube Yagi for portable work on 144MHz 28

General

Using 10GHz 34
A simple Converter for the 10m satellite band 39
An 80m Transceiver 47

Station accessories

A switched attenuator 64
An audio filter 67
A low-voltage alarm for battery supplies 70
Dry battery tester 72
A direct-reading capacitance meter 76
Earth-continuity tester 80
A diode/transistor tester 84
Extending the use of your dip oscillator 87
Dual-voltage supply from one battery 90
A simple electronic keyer 94
Field-strength measurement 97
A frame aerial for HF 99
Computer-to-radio interfaces 101
A cheap and cheerful approach to keying 105
A useful audio level indicator 108
A 'loop' alarm 110
An L-Match ATU 112

iii

A charger for nicad batteries 114
An op-amp tester 118
Optical communication 123
Packet radio principles 128
A portable power supply 132
An amplified rf probe 135
A signal injector 137
An audio-driven s-meter for dc receivers 140
A time-out unit for digital modes 145
A T-Match ATU 148
A 1750Hz toneburst for repeater access 151
A colourful voltage monitor 155
Voltage regulation 159
A bi-directional wattmeter 162
A standing-wave indicator for HF 165

Reference

Baluns 170
How the cathode-ray tube works 175
How the cathode-ray oscilloscope works 179
Diodes for protection 182
One-man holiday dxpedition 185
Getting started on a shoestring 189
A guide to HF contesting 193
Iota – a beginners' guide 204
Radio-frequency mixing explained 208
Noise-reduction circuits 210
The photometer and the polar diagram 213
The QSL bureau sub-manager's tale 216
Radiation resistance 221
A beginners' guide to RTTY contests 225
Safety, operating practice and the law 229
Screening – what is it and why is it important? 234
Speech processing 238
Your first use of a repeater 242

PREFACE

It is now some years since *Practical Projects* was first published and became popular amongst experimenters finding their way through the first intricacies of amateur radio. Together with *Radio & Electronics Cookbook* and *Amateur Radio Essentials* (both from the RSGB stable), the newcomer to amateur radio has been well served with simple projects and information to assist the learning process.

The time had come to rationalise the *Cookbook* because of rapidly-advancing technology; several of the projects in the book required components which were no longer easily available. The decision was taken to replace the *Cookbook* with another more modern volume and, at the same time, incorporate the most popular parts of *Practical Projects*. For good measure, a selection of the most appropriate articles from recent issues of *RadCom* was added.

This book is divided into two main parts: Build It Yourself; Reference. The first part is further split into: Aerials; General; Station Accessories etc.

In the 'Aerials' section, no less than six designs are there for you to try, together with two articles explaining where to erect wire aerials to best effect and how to look after them once they are up.

The 'General' section comprises three more-advanced projects; these are suitable for those people whose confidence and skill have improved to the level of tackling a more complex project.

There then follows the major section on 'Station Accessories etc'. Here you will find the best of *Cookbook* and *Practical Projects*, plus the new entries from *RadCom*. No less than 33 projects are contained in this section – something for everyone!

Finally, the second part of the book is a reference collection. Here you will find information on baluns to speech processing, on the cathode-ray tube to radiation resistance, all explained in language that no-one will find stuffy or boring.

I hope that you have several RSGB books on your bookshelf; if you have, it is a good sign that you are genuinely committed to learning about your hobby. Never discard older books – they may often contain information that the newer ones don't. Avoid being swayed by the common argument not to spend time looking backwards. Were it not for the work done in previous years, we would not be in the advanced position we are now.

I hope you will find this book stimulating and will motivate you to advance your knowledge of the subject

George Brown, MW5ACN
Knighton, Powys
2008

For the ... of Amateur ...

International Antenna Collection
Edited by Dr. George Brown, M5ACN

This book is a collection of over 50 of the very best articles published on antennas from around the world. The book is wide ranging and offers solutions to many problems experienced by the antenna enthusiast. Amongst the articles are antenna designs for most amateur bands. Stealthy and invisible antennas are covered alongside many interesting traditional designs. The book also benefits from two articles specially commissioned for inclusion here. The first, by Professor Mike Underhill, G3LHZ, of the University of Surrey at Guildford, UK, entitled 'The Truth About Loops', gives an exhaustive account of the performance of the much-maligned small loop, which also takes into account the loop's situation. approaching the same problem from the computer modelling angle is the subject of the second invited article, 'A Brief Overview of the Performance of Wire Aerials in their Operating Environments', from Jack Belrose, VE2CV. Great care has been taken to ensure that there are antennas to cover the range from 136kHz to 1.3GHz, receiving and transmitting, fixed and mobile. Everyone interested in antenna design and construction will find something in this book.

Size 275x210mm, 256 pages   ISBN 1-872309-93-3

International Antenna Collection 2
Edited by Dr. George Brown, M5ACN

This book collects together some of the best articles from around the world on the subject of antennas. It will appeal to radio amateurs in general, whether they be antenna enthusiasts or not. It is a follow-up to the successful The International Antenna Collection, compiled by the same editor. You will find antennas for most of the amateur bands. Traditional designs and highly original designs are here, simple and complex. Whatever your requirement, you will find something that is directly suited or that sets you thinking about how to solve your problem. Amongst the practical and highly erudite contents is an invited article by one of America's most respected authors on the subject of aerials. He is Kurt N Sterba, the regular 'Aerials' columnist of WorldRadio magazine. He considers one of his pet subjects - the much - misunderstood interface between transceiver and aerial. Does your ATU really tune your aerial? All the facts are clearly presented, leaving the author in no doubt as to the correct answer. As before, great care has been taken to ensure that there are antennas to cover almost all the bands between 136kHz and 2.4GHz, receiving and transmitting, mobile and fixed.

Size 275x210mm, 256 pages   ISBN 190508601-6

Each only £12.99 plus p&p

best selection Radio books

Practical Projects
Edited by Dr. George Brown, M5ACN

Packed with fifty "weekend projects" *Practical Projects* is a book of simple construction projects for the radio amateur and those just interested electronics. A wide variety of radio ideas are covered with everything from an 80m Transceiver, Antennas, ATUs and simple electronic keyers all included. Other simple electronic designs are such as dry battery testers, mobile microphones and various meters and monitors are also added. The book also contains a handy section on "now I've built it what shall I do with it?" questions answered. This book is excellent those just looking for interesting ideas to construct and for the newcomers to the hobby looking to expand their knowledge.

Size 240x174mm, 224 pages   ISBN 1-872309-88-7

Only £12.99 plus p&p

Other books by Dr. George Brown, M5ACN

Radio Society of Great Britain
Lambda House, Cranborne Road, Potters Bar, Herts, EN6 3JE Tel: 0870 904 7373 Fax: 0870 904 737

www.rsgbshop.org   RSGB SHOP

p&p on RSGB orders: £1.75 for 1 item, £3.30 for 2 or more items. Overseas rates on request.

E&OE   All prices shown plus p&p and subject to change without notice

AERIALS

Two-element, 6m Quad aerial 2
A portable three-element 6m Yagi aerial 4
Simple aerials 10
Aerial maintenance 14
A J-Pole aerial for 50MHz 17
The G3HBN portable magnetic loop 20
A UHF corner reflector aerial 25
A tube Yagi for portable work on 144MHz 28

TWO-ELEMENT, 6m QUAD AERIAL

6m – the 'magic band' – where you can work across town or try your hand at some real DX during the summer months. This simple aerial is robust – more so than a three-element beam in an exposed, windy situation, and it performs well.

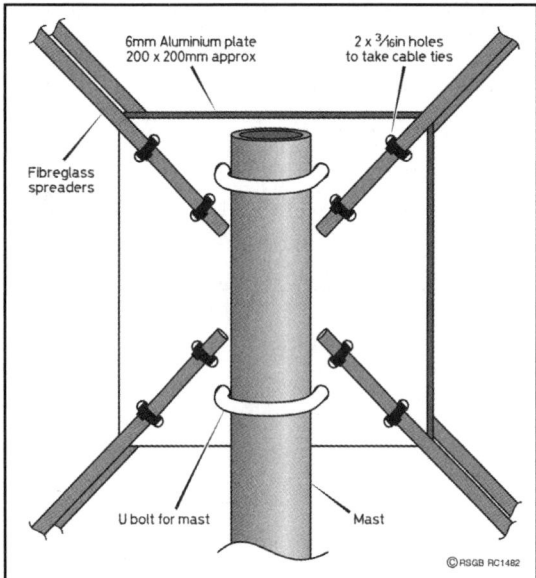

Fig 1. Glassfibre spreaders are fixed to an aluminium plate with cable ties.

CONSTRUCTION

This is simple and uses a chunk of 5mm aluminium plate, nylon cable ties and glass-fibre rods. Fig 1 details the construction.

Having drilled the metal plate and loosely fixed the rods in place with cable ties, holes are drilled about 6mm from the ends of the rods and 90° from the plane of the loop. By doing this, the wire will not slip when under tension, keeping the symmetry of the loop. The wire lengths are carefully measured, adding 30mm for soldering the ends. The loops are made up from normal multi-strand hook-up wire, although thicker wire could be used without materially affecting the performance. The driven loop and reflector have wire lengths of 543cm and 594cm respectively.

A gamma match is made up on a piece of Perspex using the earth wire out of 2.5mm mains cable for the loop. The wire is folded into a 'hairpin' 31cm long and with 2cm spacing (Fig 2). The thickness of the wire and its small size enables it to be supported only by a piece of Perspex at the capacitor and feed-point. About 20pF is required for the capacitor, so a 35 or 50pF variable is used to allow for variations.

The reflector loop is fed through the holes in the ends of the four longer rods and the two ends are soldered together. Next, the driven wire is attached to one side of the gamma match, the free ends passed through the holes of the shorter rods and the loop completed by soldering. In the final quad, the gamma match capacitor is in a small plastic box to shelter it from the rain.

TWO-ELEMENT, 6M QUAD AERIAL

Finally, the rods are slid in or out of the cable ties so the 700mm loop spacing is correct and then the cable ties are pulled tight, locking the structure in a stable, if not rigid, configuration. It can be made more rigid by tying the ends of the driven loop to the corresponding reflector loop using fishing line.

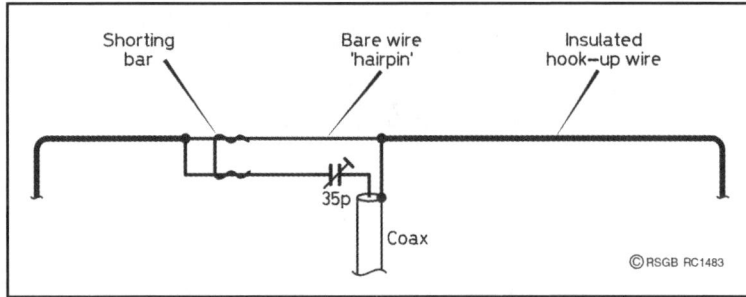

Fig 2. Details of the gamma matching system.

SETTING UP

Various feed systems were used, the best using 72Ω twin feeder, but the complexity of the balanced ATU in the shack left a lot to be desired, so the second best feed using a gamma match is described.

The gamma match consists of a coupling wire and capacitor. The tapping point and capacitor are adjusted to give a perfect match to the feeder, measured on a VSWR meter. A match can be maintained provided the feeder is kept perpendicular to the horizontal part of the driven element and equidistant to the vertical sections. During testing and setting up, you will see the VSWR rise rapidly if the feeder is moved from one side to the other, demonstrating the importance of keeping aerial systems symmetrical! **Warning! Never make any adjustments to an aerial when the transmitter is running. Always switch off the transmitter, make your adjustments, and switch on again – *Ed*.** The gamma match may be adjusted by standing on a [wooden] step ladder with the base of the quad about 8ft above ground. A shorting bar of wire is moved 5mm at a time along the hairpin, starting from the closed end, and the capacitor re-trimmed for best VSWR. This is repeated until the point of best VSWR is obtained. When mounted on the side of the house, the resonant frequency rose by 100kHz and the VSWR rose to 1.2:1 due to the change in location.

The quad aerial is easy to construct if the right materials are at hand and it is easy to maintain. The glass-fibre rods used were spares, but they are available from garden centres for making cloches. Failing this, bamboo canes could be used, but these must be selected for thickness and flexibility so that the final shape is satisfactory.

PARTS LIST

Glass-fibre rods, 8 or 10mm diameter, 4 × 108cm long and 4 × 111cm long
Aluminium plate, 20 × 30 × 0.5cm
2 × 50mm U-bolts. Car exhaust clamps are ideal
12 × 5mm nylon cable ties. Make sure you have spares
Variable capacitor, 50pF
Copper wire, 1m solid, 2.5mm
Multi-strand hook-up wire, 12m

A PORTABLE THREE-ELEMENT 6m YAGI AERIAL

So, you've just acquired a transceiver with HF and 6m coverage, and you are keen to try out the 'magic band' for the first time. What better way than with a simple, portable antenna, which you can dismantle and use for some /P operation during the summer months?

REQUIREMENTS
When contemplating the addition of another aerial, most people would list their requirements as follows:

- Boom length, maximum 2m
- Reasonable gain
- Low cost
- Ability to dismantle easily for transportation
- Easy assembly with the minimum of tools
- Easy and quick tune-up
- Lightweight

First, materials are needed. A visit to the local DIY shop should provide a good source with a wide range of 1m and 2m lengths of aluminium tubing and box sections. Use 12.5mm diameter tube in 2m lengths for the centre sections of the elements, and 10mm for the element end sections. These sections of tube fit snugly. You will also need a 1m length of 10mm rod to make the gamma match section. A 2m length of 25.4mm square U-section aluminium was used for the boom.

THE DESIGN
A three-element beam always has one reflector, one driven element and one director. For simplicity, 'plumbers delight' construction is employed so all the elements and the boom are at a common earth potential. This reduces some of the static electricity which can be prevalent in other types of design. However, this limits the feeding arrangements to a delta or a gamma match. The gamma match was chosen because of its easy adjustment, bearing in mind the requirements for quick tune-up and transportability.

DRILLING THE BOOM
The first thing to do is to measure out the boom and drill the holes for the reflector and director elements (see Fig 1 for dimensions). The 12.5mm holes are marked and drilled with a smaller (6mm) drill bit before the larger holes are drilled. Care must be taken to get these holes straight and level in both directions, otherwise the aerial will look very odd indeed. Using a small round file, file a notch in the top side of each of the four holes. The notch should be just large enough

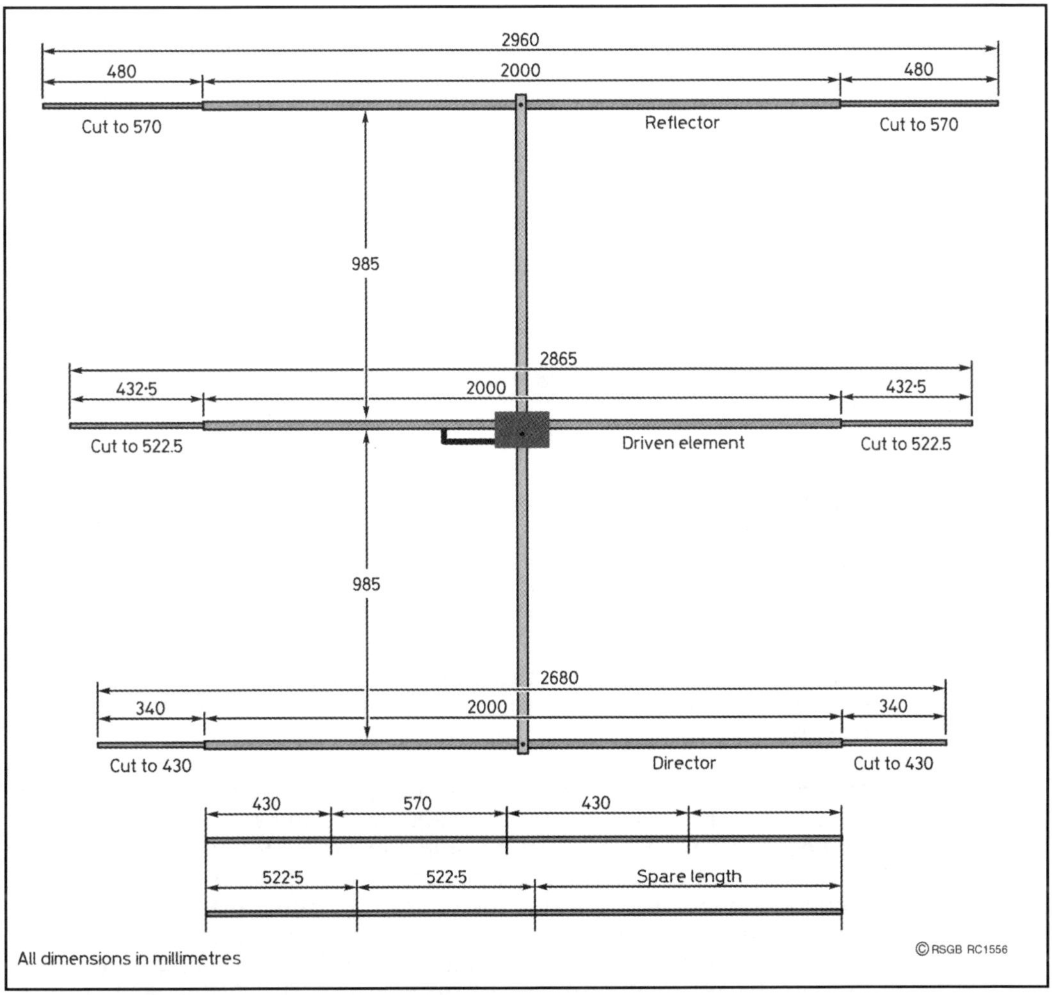

A PORTABLE THREE-ELEMENT 6m YAGI ANTENNA

Fig 1. Dimensions and tube-cutting

to pass the head of the screw which holds the 10mm tubing to the 12.5mm tubing (see Fig 2). The reason for doing this is simple; it makes it easier to dismantle or assemble the elements on the boom. In the flat side of the boom drill a 2mm hole on the element centre line; this is for the screws to hold the elements in place (see Fig 2).

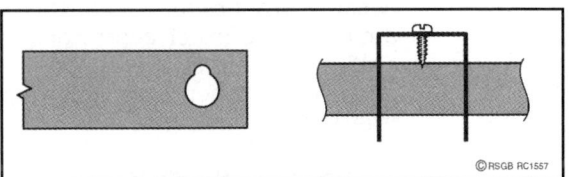

Fig 2. Details of the element holes (reflector and director).

DIRECTOR AND REFLECTOR
The 'standard' sizes of tubing available are perfect for a 6m beam. I bought two 2m lengths of 10mm tube and carefully cut them to the dimensions shown in Fig 1. In this way one piece gave me the correct lengths for four end-pieces (reflector and director). With these end-

WEEKEND PROJECTS

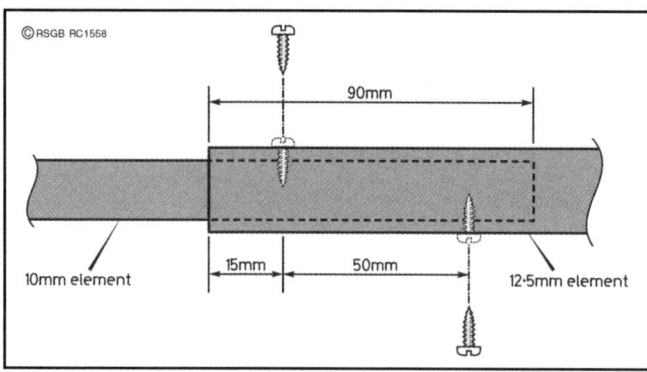

Fig 3. Detail of joining end-pieces to the main elements.

pieces cut to the right lengths, they were simply inserted into the 2m lengths of 12.5mm tubing and, after marking the correct length of the whole element, 2mm holes were drilled for the self-tapping screws to be driven through both the 10mm and 12.5mm tubing to hold them together. Each screw was driven through the tubing in a similar manner on opposite sides. This is done on the director as well as the reflector (see Fig 3).

DRIVEN ELEMENT

The driven element is made of the same materials as the reflector and director. It is fixed to the top of the boom together with a plastic weatherproof box, which houses the 50pF capacitor and the gamma arm assembly.

The fixing screw through the box into the boom allows the driven element 12.5mm tubing to go in parallel to the boom for easy transportation.

At this stage, the 10mm tubing and the 12.5mm tubing are not yet fixed together. This is so that the 12.5 mm tubing will fit tightly into the connections box. Holes must be drilled in the 12.5 tubing of the driven element for the gamma arm. These holes allow connection of the gamma arm by a screw inside the tubing. A larger hole (8mm) is drilled through one wall of the tubing, and a 2mm hole is drilled in the opposite wall. The reason for the 8mm hole is to allow a screwdriver access to tighten the element-to-gamma arm screw. After fitting and testing, the 8mm hole can be filled with putty or simply taped over with weatherproof tape. To get the correct location of the gamma arm holes, first find the exact centre of the driven element and then measure out 305mm – this is the centre-line for the holes. At the centre of the element, another 2mm hole must be drilled for the shield connection of the coaxial cable.

CONNECTIONS BOX

The connections box needs to be drilled to take the driven element, gamma matching arm and the capacitor shaft. The first holes to mark and drill are for the driven element. These were carefully marked on each end of the bottom half of the plastic box, then drilled with a 6mm pilot drill and then again with the 12.5mm bit. The gamma arm hole is drilled 40mm away from the element hole, and is only drilled in one end of the box. The driven element and 1m aluminium rod are placed into the box and a suitable position for the capacitor located and marked for drilling. The capacitor came from the junk box, and the exact position is not critical as long as it does not foul the element or gamma arm

A PORTABLE THREE-ELEMENT 6m YAGI ANTENNA

(see Fig 4). A hole is required to screw the connections box to the boom; this is a 2mm hole drilled in the centre of the box. This screw goes through the plastic box and into the boom from the top side. The other hole required at this time is one for the coaxial cable. This is drilled in the same end as the gamma assembly hole in the centre of the box.

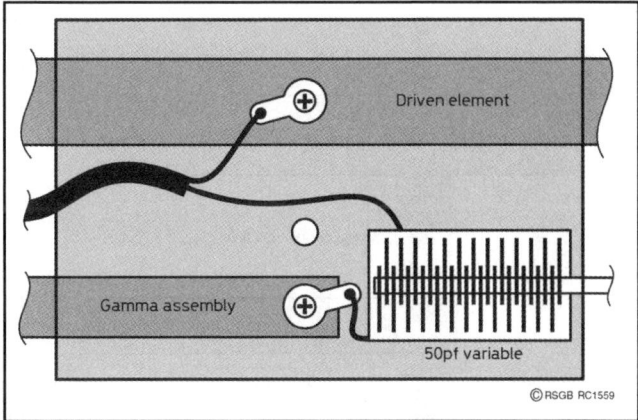

Fig 4. Connections from the coax to the gamma match.

GAMMA ARM

With the 1m length of aluminium rod, measure and cut 320mm off. At one end, drill a 2mm hole for the connection point of the capacitor. 90° out from this, at the other end of the rod, drill a 2mm hole straight through the rod. Using a 6mm drill bit, drill half way through the rod to make a countersunk hole in the rod (see Fig 5). This is the attachment point for the gamma connecting arm.

From the left-over length of aluminium rod cut a piece exactly 40mm long. With a file, one end must be filed to fit the gamma arm, and the other end filed to fit the 12.5mm tubing.

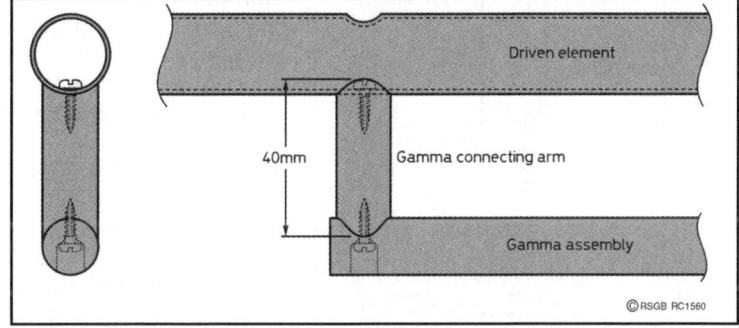

Fig 5. The gamma arm assembly.

If a drill-press is available, it can be used to make a perfect job by drilling through the rod with the correct size of drill bit. However, the file also works well, though it does take a bit longer. In each end of the gamma connecting arm a 2mm hole is required for the fixing screws (see Fig 5).

Using a self-tapping screw, screw the gamma connecting arm to the gamma arm, forming an 'L'-shaped gamma assembly. Place a screw through the 8mm hole in the driven element and screw the gamma assembly to the element using a self-tapping screw (Fig 6).

Slide the connections box over the driven element with the hole for the gamma assembly facing the gamma assembly. Insert the gamma assembly into the hole in the box and adjust the box until it is in the centre of the element. Now screw the end-pieces into the driven element in the same manner used for the director and reflector (see Figs 4 and 5). Fit the capacitor to the box and connect one side of the variable capacitor to the gamma assembly, using the 2mm hole (drilled

WEEKEND PROJECTS

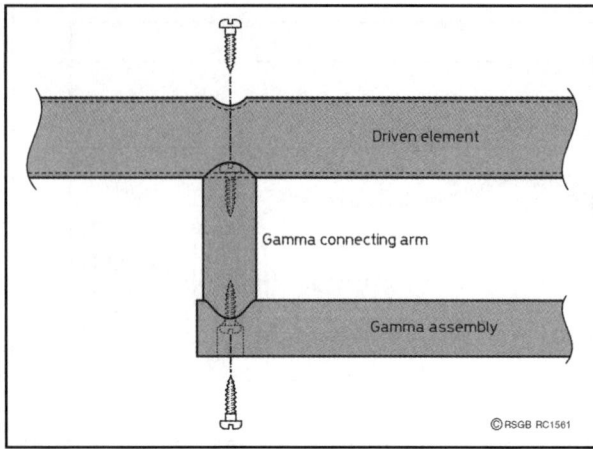

Fig 6. Connecting the gamma arm assembly to the element.

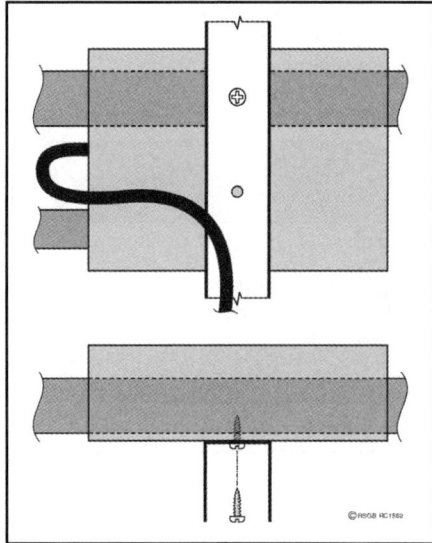

Fig 7. Connecting the box to the boom.

earlier) and a solder tag held in place with a self-tapping screw. Feed the coaxial cable into the box from the outside and strip the ends ready for connection. The shield of the coaxial cable is soldered to a solder tag and screwed to the driven element using a self-tapping screw. The inner of the coaxial cable is soldered directly to the other side of the variable capacitor. All the external connections to the box can now be weatherproofed using hot-melt glue, epoxy or similar product.

FIXING THE DRIVEN ELEMENT

Locate the centre of the boom and drill a 2mm hole in the exact centre of the flat edge (see Fig 7). The connections box is screwed to this point using a self-tapping screw. Now, turn over the boom and align the driven element so that the boom and the element are at 90° to each other. Drill through the boom and the plastic box and into the driven element with a 2mm drill bit. This hole will be off-centre and it holds the driven element in the correct place when the beam is being used.

WEATHERPROOFING

Fill the ends of all the elements with hot-melt glue, epoxy, or a similar long-lasting product to keep rain out of the tubing. After checking all the connections are good in the connections box, close the lid and seal against the weather. If required, drill holes in the side of the boom for a mast clamp.

CHECKING THE SWR

The VSWR should be quite good if the dimensions in this article are followed closely. Place the assembled aerial in a clear area at least 3m off the ground and check the VSWR on a known VSWR bridge. The VSWR can be adjusted by turning the capacitor and checking again. It is best to check the VSWR at both band edges and set the VSWR minimum at the centre of the band (see Fig 8 for VSWR readings on the three beams built for testing purposes).

ASSEMBLING AND DISMANTLING

Lay out all the metal pieces in a clear flat area. Rotate the driven element to 90° and insert the retaining screw through the boom and into the driven element. Slide the director through the boom and lock it in place using a screw through the boom and into the element. Do

A PORTABLE THREE-ELEMENT 6m YAGI ANTENNA

the same thing with the reflector. The complete aerial is shown in Fig 9.

To dismantle the beam, lower the mast, remove the aerial and the three screws in the boom. Remove the director and reflector and rotate the driven element through 90°. Only a single screwdriver is required for the beam and three screws.

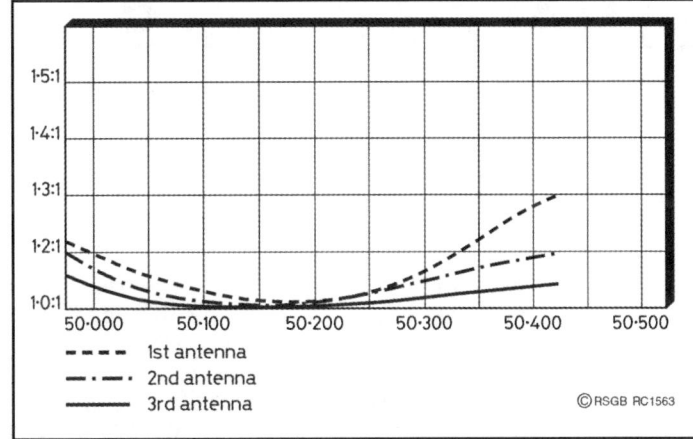

Fig 8. VSWR readings on the three beams built for testing purposes.

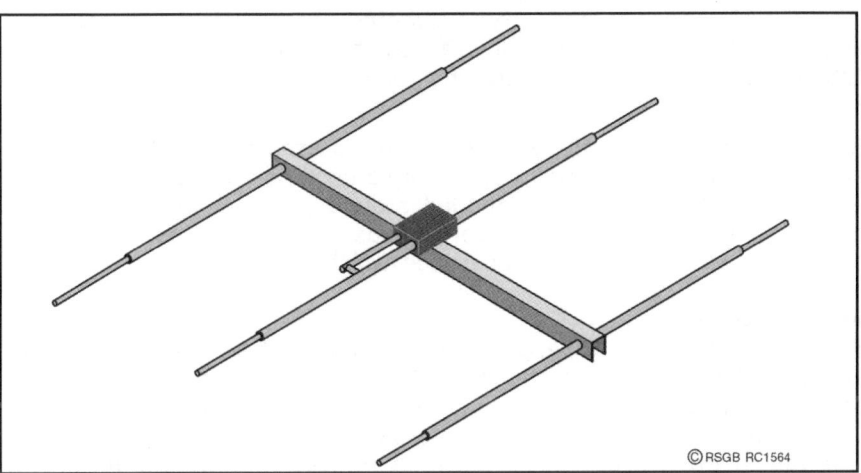

Fig 9. The complete three-element 50MHz beam.

CONCLUSIONS

The three-element beam meets all the criteria which were set out at the beginning. It has proved robust enough for everyday use at home as well as portable use on my 10m pump-up mast. The total weight for the aerial is less than 3kg, and its total cost £18.63.

PARTS LIST

Boom section	One 2m length, 25.4 × 25.5mm square 'U' aluminium
Element centres	Three 2m lengths 12.5mm OD aluminium tubing
Element ends	Two 2m lengths 10mm OD aluminium tubing
Gamma match	One 1m length 10mm aluminium rod
Connections box	Plastic box 70 × 122 × 50mm
VC1	Variable capacitor, 50pF
Other items	Solder tags, self-tapping screws

SIMPLE AERIALS

The performance of any receiver or transceiver, no matter how expensive it is, is limited by the aerial that feeds it. Two of the most frequently asked questions are:

- Which is the best sort of aerial to use?
- Where is the best place to locate an amateur radio aerial?

To answer these questions, you must ask yourself what sort of operation you want to do. Are you interested in local, chatty contacts on the lower bands or VHF, or are you more disposed towards long-distance (DX) contacts, and on what band?

A house with a moderately-sized garden is assumed in the diagrams here, to illustrate the configurations of some simple aerials. You would not need all these aerials festooned around your house, because one or two would be sufficient for your needs. The problems incurred by properties with more restricted space will be/covered later.

VHF AERIALS
For VHF operation, the aerial should be mounted as high as possible, either on a mast or on a chimney. For all-round coverage on FM and the local repeaters, a vertical colinear is a good choice. For SSB and CW DX operation, a horizontal rotatable beam is needed. If satellite working is envisaged, you will need to contemplate mounting an elevator on top of your rotator, so that your beam can point in any direction, including vertically upwards! An advantage of satellite working is that the aerials do not necessarily have to be up in the air, provided you have a relatively uncluttered site. Your rotator and elevator can be at ground level, which is good!

If the VHF aerial is mounted on the chimney, use a double mounting bracket, particularly if you have a beam and rotator. Keep the TV, broadcast FM and amateur aerials as far apart as possible, and keeping the feeders separated is also a good plan.

THE DIPOLE AERIAL
One of the simplest types of aerial for single-band operation is the half-wave dipole. (The name 'dipole' simply means

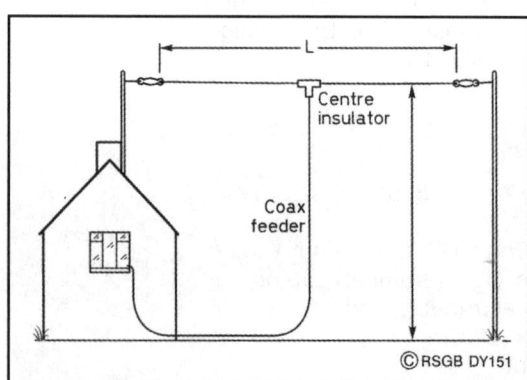

Fig 1. Layout for a dipole aerial.

SIMPLE AERIALS

'two poles' or 'two elements', and in this case the total length of the dipole is approximately half a wavelength at the operating frequency.) It is usually fed in the centre by coaxial cable as shown in Fig 1. The length of the dipole for the lowest frequency in each band is shown in Table 1. Normally, the length of the aerial will be 'trimmed' to be tuned to the centre frequency of the part of the band in which you will operate. This is done using the data in the right-hand column of Table 1. As an example, suppose you wanted your aerial to be resonant at 3.7MHz. The table gives an overall dipole length of 42.86m for 3.5MHz. To resonate the aerial 200kHz higher, then this length must be shortened by 2 x 0.595 m = 1.190m. Your dipole would thus be 41.67m long. Remember to allow extra wire for fixing the dipole ends to insulators.

Band (MHz)	Dipole length (m)	Trim each end (mm/l0kHz)
1.8	83.33	2190
3.5	42.86	595
7	21.43	150
10	14.85	70
14	10.71	35
18	8.33	20
21	7.14	15
24	6.03	12
28	5.36	10
50	3.00	6

Table 1. Dipole lengths for lowest frequency of each band and the length to be trimmed from each to raise the resonant frequency by 100kHz.

On the lower-frequency bands, the lengths become rather large. In this case, you can 'bend' your dipole, as illustrated in Fig 2. The length of wire required to give an acceptable value of SWR (less than 2:1 on transmit) may need to be different from the calculated value, so be prepared to experiment!

Dipoles are single-band aerials, although they will often work acceptably on the third harmonic of their design frequency – a 7MHz dipole often operates reasonably well on 21MHz. It is possible to operate several dipoles in parallel, as Fig 3 shows. Interaction between the elements can occur if the spacing between them is less than about 10cm. A multi-band dipole, as shown in Fig 3, has the elements separated with plastic spacers, and drooping ends to produce maximum spacing between the elements' ends.

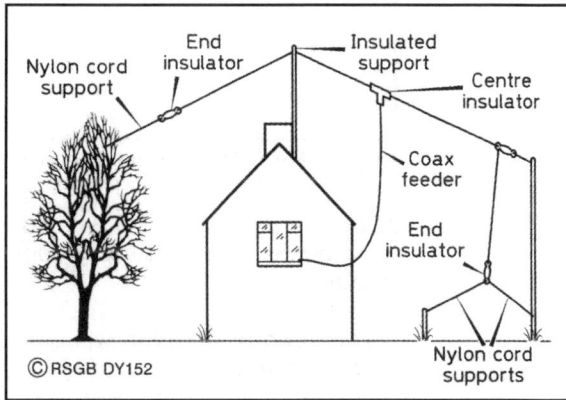

Fig 2. Possible layout for a dipole aerial in a confined space.

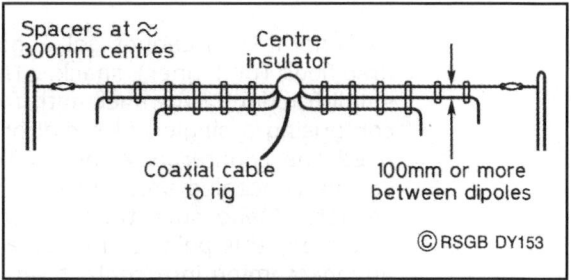

Fig 3. Multi-band dipole aerial.

WEEKEND PROJECTS

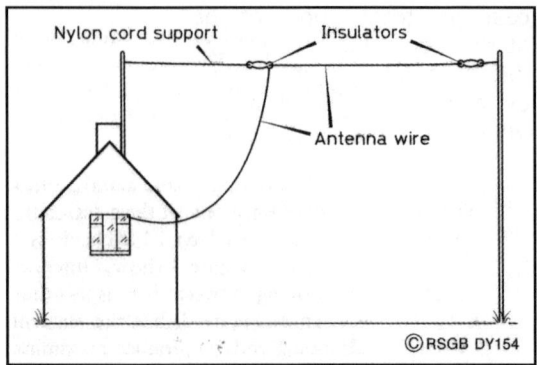

Fig 4. Long-wire or inverted-L aerial.

THE LONG-WIRE AERIAL

This aerial is simple, cheap, easy to erect, and suits most houses and gardens, as Fig 4 shows. Using an aerial tuning unit (ATU), an end-fed long wire can function on several bands when used with a set of radials or a counterpoise. Fig 5 illustrates the setup. The length of the aerial will determine the bands which will be covered.

A wire length of 10.5m will work on the 40, 30, 17, 15 and 12m bands.

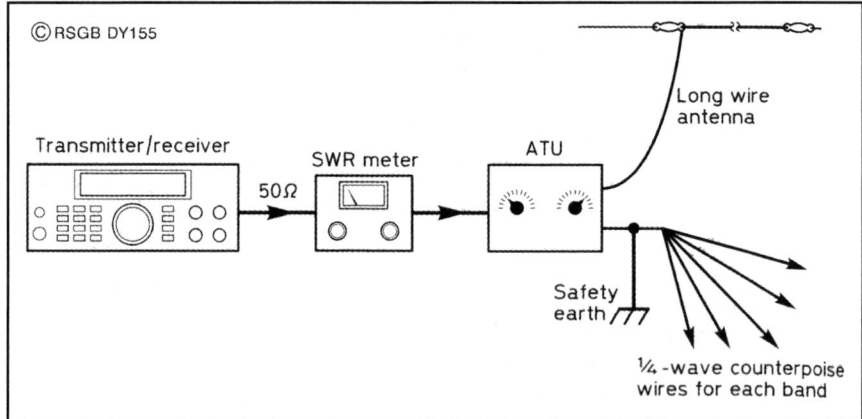

Fig 5. How to connect the radio to a long-wire aerial.

A wire length of 15.5m will work (with an ATU) on the 80, 40, 20 and 12m bands and possibly (depending on your ATU) on the 17 and 15m bands.

A wire length of 26.5m will operate on all bands, but may be difficult to load on 10m.

The wire lengths given here may need some adjustment because of the geometry of your particular house and garden. For receive-only purposes, the lengths are far less critical.

In general, you cannot get a good radio-frequency (RF) earth from a first-floor (or higher) shack. Unless a good RF earth exists within a small fraction of a wavelength from the transceiver, an artificial ground comprising a single $\lambda/4$ radial or counterpoise will be needed. You will need one counterpoise for each band you intend to use, and the wire can be concealed around the skirting-board of the shack, or under the carpet. Make sure that the free end of each counterpoise is well insulated; this point can carry a very high voltage when you transmit; anyone coming into contact with this can suffer very severe RF burns. Counterpoise lengths can be read from Table 2.

SIMPLE AERIALS

Band (MHz)	Element length (m)
1.8	39.66
3.5	20.40
7	10.20
10	7.14
14	5.1
18	3.96
21	3.4
24	2.99
28	2.55

Table 2. lengths of elements for vertical antennas, radials for verticals and counterpoises for end-fed long-wire antennas.

THE VERTICAL AERIAL
The single-band vertical aerial is sometimes used by DX operators because it has a low angle of radiation, which favours long-distance propagation. However, it must be sited clear of obstructions and must have a good counterpoise or radial system. Illustrations of the vertical aerial are shown, and the lengths of the vertical and radial sections are given in Table 2. The centre of the coaxial feeder is connected to the vertical section, and the braid to the counterpoise or radial system, which is made up of four or more wires buried just below the surface and joined together near the base of the aerial.

CABLE ENTRY TO THE HOUSE
Bringing coaxial cable into the house by an open window must be regarded as a temporary measure. Wooden window frames can be drilled, one hole for each feeder. Make the holes slope downwards from inside to out to prevent rain entering, and treat these with wood preservative. Leads from long-wire and inverted-L aerials should be kept separate from other cables.

Alternatively, a plastic pipe large enough to take all your feeders could be fitted into the brickwork (again, sloping downwards towards the outside). You may want to let a friendly builder do this for you.

AERIAL MAINTENANCE

Spring is the ideal time to do something about those aerials that have suffered during the extremes of the winter months. This work involves maintenance of existing aerials and the installation of better ones.

AERIAL UPKEEP
We all know that a length of wire will radiate but a length of wire will radiate even better if it is in good condition. Most aerial losses are caused by corrosion, which increases the resistive losses in the aerial. When maintenance is carried out on an aerial system, all mechanical joints should be dismantled and corrosion removed. They can then be coated with grease before reassembly. Soldered joints should be inspected and remade if they look suspect. Insulators, particularly at high voltage points (such as at the ends of dipoles or long wire aerials), should be cleaned.

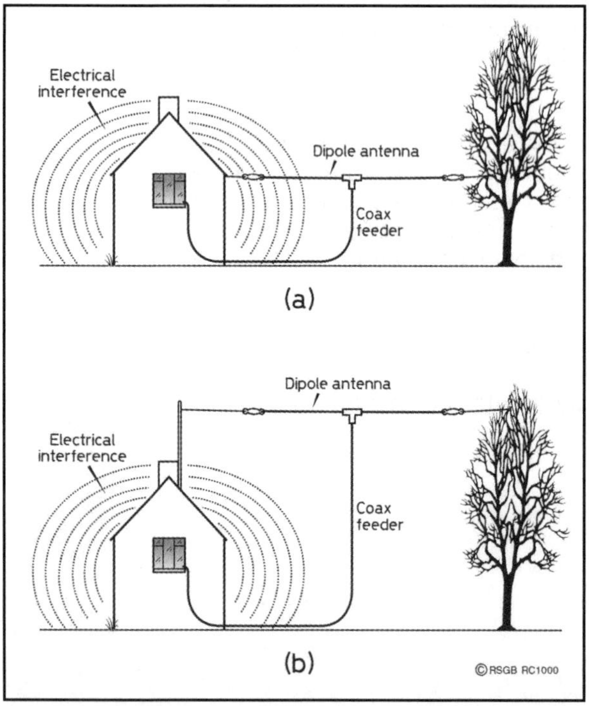

Check that the transmission lines (coaxial cables or twin feeders) are in good condition and that the connectors are free from corrosion. They, too, should be coated in grease after cleaning and before re-assembly.

IMPROVING THE AERIAL
In general, the most practical way of improving the performance of an existing aerial is to erect it as high as is practicable at your location. An aerial that is low and close to the house is also close to the electrical QRM which envelopes our dwellings as shown in Fig 1(a). If the aerial can be raised it will be further from this interference, which will enable us to hear stations that would otherwise be hidden in the electrical noise; see Fig 1(b).

Fig 1. (a) Dipole aerial at a low height, showing it in the strong electrical interference field; (b) aerial height is increased, thus reducing the electrical interference and increasing the signal.

The aerial will also be further away from TV sets and other domestic electronic equipment and the chances of TVI are reduced. Also,

AERIAL MAINTENANCE

when the height of the aerial is increased, the radiation pattern favours DX stations to a greater degree (because of the lower angle of radiation).

A worthwhile improvement can often be made with a modest increase in height, such as moving one end of the aerial from the eves of the house to the chimney or connecting it to a higher branch of a tree.

TRANSMISSION LINES (FEEDERS)

Although the aerial is installed as high as possible, the transceiver is installed at a convenient place indoors, where it is readily accessible and out of the weather.

For the best signal-to-noise ratio, and for avoiding TVI, it is better if the connector between the aerial and the transceiver does not radiate or receive signals. This is achieved using a transmission line, which can comprise coaxial cable ('coax') or twin-line feeder.

The twin-feeder form of transmission line comprises two conductors placed parallel and close together as shown in Fig 2(a). The current flowing in the two conductors travels in opposite directions; in other words they are 180° out of phase with each other. If the two currents also have equal amplitudes, the electromagnetic field generated by each conductor will cancel that generated by the other, and the line will not radiate or receive radio energy.

With coax (Fig 2(b)), the current flowing in the outer conductor does so on the inner surface so that radiated or received RF energy is cancelled, just the same as in the twin-line feeder. In addition, the outer conductor of coax acts as a shield, confining the RF energy within the line. Because of its construction, coaxial cable is said to be *unbalanced*.

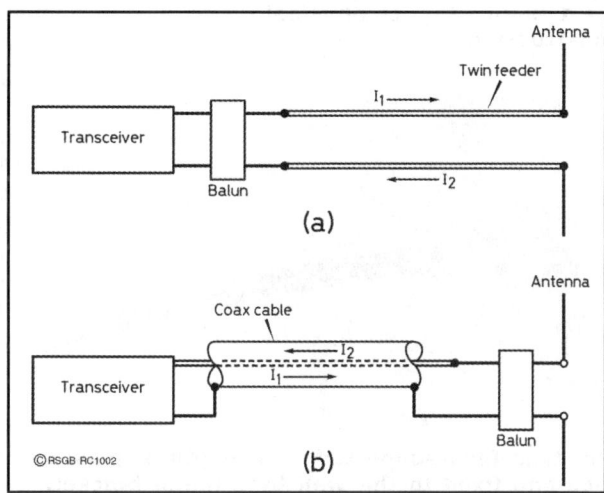

If the currents flowing in the two lines of the feeder are not equal in amplitude or not exactly 180° out of phase, the line will radiate or receive RF energy.

Fig 2. Showing the field-cancelling effects of current in (a) twin feeder and (b) coaxial cable.

Centre-fed dipoles and loops are *balanced*, meaning that they are electrically symmetrical with respect to the feed-point (where the feeder is connected to the aerial). A balanced aerial should be fed with a balanced feeder system to preserve this electrical symmetry with respect to ground. However, the aerial connector on the back of the transceiver is for coaxial cable, and is thus unbalanced. Some method of method of connecting the transmission line to the aerial without

WEEKEND PROJECTS

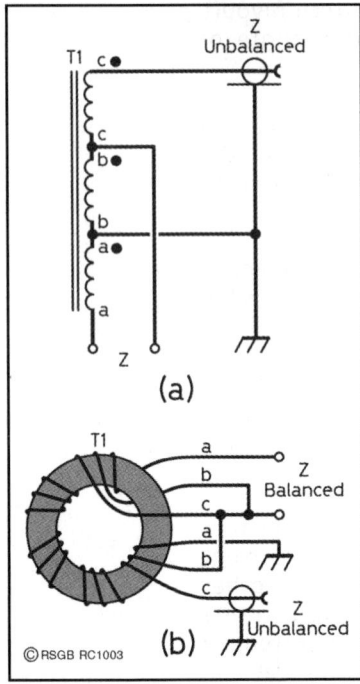

Fig 3. (a) Circuit diagram of a 1:1 balun; (b) its physical construction.

upsetting the symmetry of the aerial itself is required. A device for converting a balanced circuit to an unbalanced circuit is a *balun* (a contraction of 'balanced to unbalanced'); see Fig 3.

Coaxial cable has the advantage of being very practical for most amateur radio installations. Because of the excellent shielding afforded by its outer screening, coax can be run up a metal tower or taped together with numerous other cables with virtually no interaction. At the top of a tower, coax can be used with a rotating beam without shorting or twisting conductor problems. Coax can even be buried underground in plastic tubing.

The disadvantage of it is that it normally requires some matching unit or balun at the aerial. Also, precautions have to be taken to ensure that the connection is absolutely watertight. If water gets into the tube-like structure of coax then the cable is ruined. Coaxial cable is relatively heavy compared with a single copper wire. This means that there is a fair amount of mechanical stress on the coax-to-aerial connection of a centre-fed dipole, especially when using heavy-duty coax.

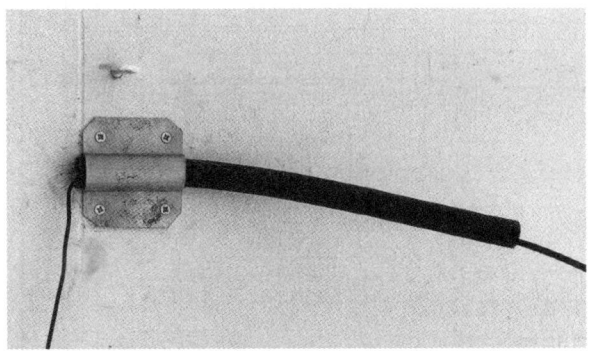

75Ω twin-feed supported in a length of plastic tube and fixed to the wall by a metal bracket.

Open-wire feeder must be kept away from metal objects by several times the spacing between its conductors. Despite this mechanical difficulty, there are often compelling reasons for using this type of feeder. One of these is the low loss incurred when using twin-line feeder in a multi-band aerial. In addition, twin feeder is far cheaper than coaxial cable and it is much lighter.

The 72Ω twin feeding my quad is now six years old and has not yet failed. I used to change my coax each spring!

An excellent and cheap insulator for HF aerials is a 1m long length of heavy monofilament fishing line or strimmer cord, but learn to tie a bowline or fisherman's knot otherwise it will easily come undone!

A J-POLE AERIAL FOR 50MHz

For FM communication (ie voice and data) on the VHF bands, a vertical aerial is used to give all-round (non-directional) coverage. This is a half-wave aerial which can be fed at the end, thus removing the principal problem with the conventional vertical centre-fed half-wave dipole, which is that the feeder should leave the dipole at right angles. This is no problem when the dipole is horizontal, but can be difficult for the vertical dipole.

BASIC FACTS

A feeder must be connected to an aerial at a point where the impedance (AC 'resistance') of the aerial closely matches that of the feeder. The difference between the two impedances gives rise to the voltage standing-wave ratio (VSWR), which is unity only when the two impedances are the same. With 50Ω feeders, the feed-point of a half-wave aerial is at the centre, where the aeral impedance is around the same value. At the end of a half-wave aerial, the impedance is high, so it is not a suitable point to connect a 50Ω feeder.

Connection at this point can be effected using an RF transformer. RF transformers act in the same way as ordinary transformers, except that they are much smaller, and usually comprise wires of particular lengths adjacent to each other. Fig 1 is a good starting point. It shows the aerial in its diagrammatic form. Notice that the aerial is in the form of an elongated letter 'J'; this shape gives rise to its nickname – the J-pole. The quarter-wave RF transformer is the lower 'U'-section below the half-wave element. At the bottom of the U-section, the impedance is zero (this may become clearer later) and at the top of the U-section it is high, thus matching the aerial impedance. The coaxial feeder cable is connected part-way up the U-section, where the impedance is around 50Ω.

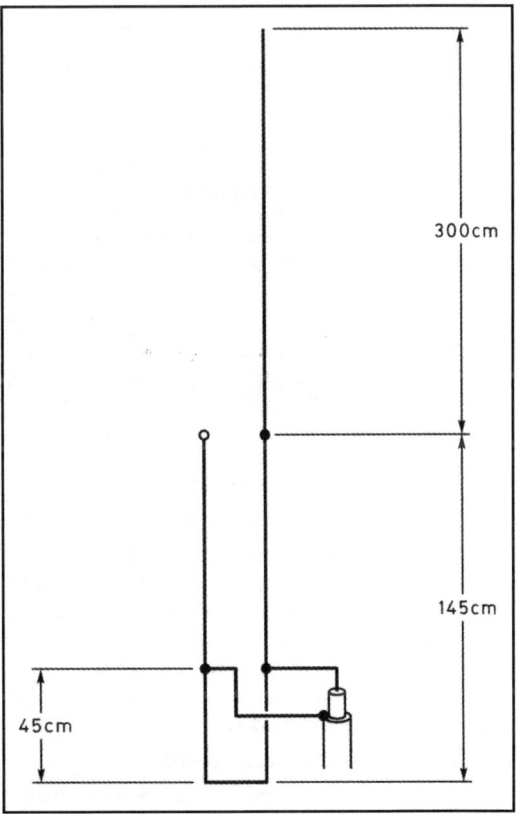

Fig 1. Overall dimensions of the 6m J-pole aerial.

WEEKEND PROJECTS

THE PRACTICALITIES

Fig 2 shows the aerial as constructed. As it is about 4.5m high, it may be too high for the average house loft, but is ideal for mounting outside, supported by a non-metallic pole or hung from a tree branch. The upper half-wave section is made from 1.5mm insulated copper wire, as used in domestic mains wiring. The quarter-wave transformer below the half-wave section is made from 300Ω balanced line ('ribbon cable'). The wires at the bottom of the transformer section are stripped of their insulation, twisted and soldered. At the upper end, only one wire of the balanced line is soldered to the bottom of the half-wave section. The other wire of the pair is not connected and is left insulated.

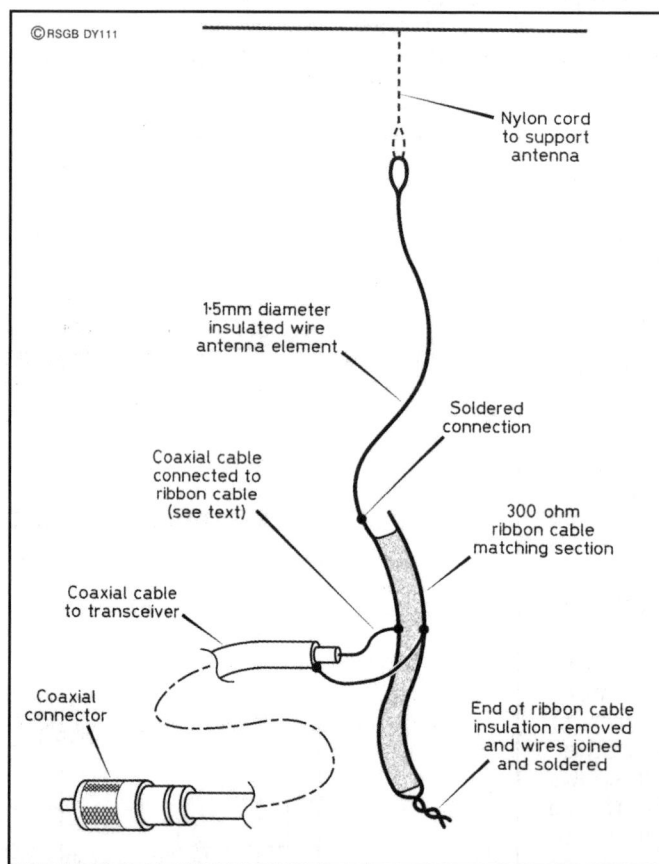

Fig 2. Constructional details.

At the feed-point of the transformer, the insulation needs to be carefully stripped from the balanced line. You will need a standing-wave meter (VSWR meter) in the coaxial line between your transmitter and the aerial, and you will need to adjust the position of the feed-point; 45cm from the bottom was the best point on the prototype, but this position is dependent upon the immediate surroundings of the aerial, and must be done when the aerial is in its final operating position. **Warning: Never make adjustments to the feed-point when the transmitter is on.** Make a VSWR measurement, switch off, move the feed-point, switch on again, make another measurement, and so on. You will need to aim for the lowest VSWR; you can certainly get better than 2:1. Having found the best position, wrap all the exposed wires with self-amalgamating tape, to seal them against the ingress of moisture.

HOW IT PERFORMS

Fig 3 shows a computer prediction of how the J-pole radiates. It is called a polar diagram, and shows the distribution of your transmitted power when viewed 'from the end of your garden'. Most of your signal is sent at a fairly small angle to the horizontal; very little signal goes

A J-POLE AERIAL FOR 50MHz

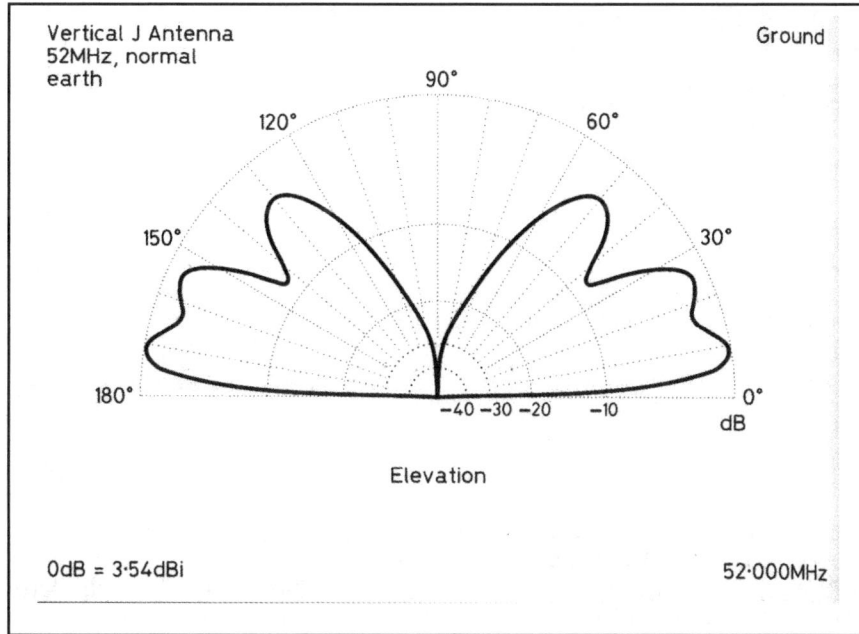

Fig 3. The computed polar diagram of the J-pole.

upwards, which is a good thing, of course. This also shows why the J-pole (or any other vertical aerial) should not be called 'omnidirectional', which means it radiates in all directions. It is omnidirectional only in the horizontal plane.

SAFETY
Where you mount your aerial is a matter of personal preference and the restrictions of height and space, but the following safety rules must be applied.

1. Never fix an aerial where it may come into contact with power lines or telephone lines.
2. When climbing a ladder to put up an aerial outside, make sure the ladder is safe and that it is secured.
3. Don't do this alone. Preferably have someone with you. If this is not possible, make sure someone knows where you are.

PARTS LIST
3.00m	1.5mm insulated copper wire
1.50m	300Ω balanced line
As required	50Ω coaxial cable
As required	Self-amalgamating tape

THE G3HBN PORTABLE MAGNETIC LOOP

For many years, a portable magnetic loop has been in use at G3HBN for holidays and special events. The design was a simple loop of RG-213 braiding slid over a piece of half-inch water hosepipe and supported by pieces of bamboo which formed a pear-shape. The whole was manually tuned and supported on a photographic tripod. It was time to upgrade this loop. Some articles in *RadCom* inspired a new design.

The requirement was to improve the performance and increase the operating bandwidth, if possible. Looking at some aerial history, the old Cage Dipole came to mind. The Cage Dipole was designed to increase the bandwidth and help with the matching of it for commercial broadcasting purposes. In those stations, very long open-wire feeders from the transmitter to the aerial were customary. Aerial tuning units were not used and the feeder was coupled directly into the transmitter with either a link or a π-coupling circuit. Such aerials would have a bandwidth of say 2.5 to 5.0 or 5.0 to 10.0MHz.

THE LOOP

The element of this loop is constructed along the lines of the cage dipole outlined above, but only a single cage element is used. This element consists of 12 cables of stranded plastic-covered hook-up wire connected in parallel. The inner core of wire is about 1mm diameter. The overall diameter of the cage is 50mm. The element, when constructed, is placed on an hexagonal wooden frame. A ganged tuning capacitor of 525 + 525pF with slow-motion drive is used to bring the loop to resonance at the desired frequency. The tuning range is from 6.9 to 32.0MHz. The loop is fed with a Faraday link coupling made with RG-213 coax, the braiding of which is open for 2.5cm at the centre. The photograph shows the completed loop on a tripod mounting. One of the problems of magnetic loops is the very narrow bandwidth. Table 1 illustrates the comparison between a 1m, 22mm copper tube element and the 1m, 50mm caged element.

Portable loop 12-wire cage	Centre Frequency	Fixed loop 22m copper tube
21	7015	11
30	10115	16
50	14050	30
75	18100	45
110	21100	60

Notes: All values in kHz.
Measurements were taken at VSWR of 1.3:1 points with a Welz SP-300 VSWR / power meter.
Both loops had a VSWR of 1:1 at the centre frequency.

Table 1. Bandwidths of the two designs.

THE G3HBN PORTABLE MAGNETIC LOOP

A worthwhile bandwidth increase has been achieved. This represents an overall improvement in performance and is reflected in the results. The outer diameter of the loop at the diagonals is about 105cm. When the length of the conducting element becomes greater than $\lambda/4$, the loop ceases to operate properly and becomes difficult to couple. It is desirable, therefore, to try to keep the overall conductor length to about 0.24λ or less at the highest frequency of operation. Although the model described here will operate on 29MHz, the element is really just a little too long.

CONSTRUCTION

Most constructors seldom follow exactly what is described in an article but, listed in Table 2, are the items that went into making this particular model.

3	old CDs glued together to form the centre core
2	plastic 3cm plumbing nuts glued together
7	plastic till-roll spools or similar rigid plastic tube that fits the dowels
12mm	dowelling for the 6 spokes of the hexagon
5	plastic 10mm or 15mm wall-mounting water pipe clips
4	2.5cm (1in) rubber tap washers
5	3 x 40mm bolts with nuts and washers
5	water pipe saddles to strengthen centre mountings
12	55mm plastic discs.
4	packets of 10m of 6A (24/0.2mm) hook-up wire (Maplin)
	Connectors to suit termination to the capacitor
1	525pF + 525pF variable capacitor
1	suitable plastic box or container
1	slow motion drive 6 or 7:1 reduction.
1	well-insulated tuning knob, or plastic coupler and knob
1m	RG-213 Coax
5	Terry clips, 10mm
	Portable and rotary mounting methods

Table 2. Parts list.

The assembly is fairly obvious from the photographs and Fig 1. There are several points that are not so obvious. The overall diagonal measurements from the centre to the outer cables should not be less than about 105cm. If the hexagon is smaller than this, with the cable specified, the loop might not quite tune to 7MHz. With this measurement, the loop should tune from 6960kHz to 32MHz.

The dowelling should be cut into five lengths of 47cm and inserted into rigid plastic tubes (eg till-roll inners) at the centre hub. The sixth length is measured to fit whatever mounting box can be found. The seventh plastic support should be mounted to the tuning box and then the sixth spoke measured and cut. The five white plastic pipe clips are screwed to the ends of the five 47cm spokes.

The spacers are made with the 12 plastic discs cut from about 1mm thick plastic. A 3-litre food container is cut into flat pieces which are

WEEKEND PROJECTS

The loop in use at G3HBN.

scored with 55mm circles. A further circle of 50mm diameter is scored inside each one. Marks should be made every 30° on the 50mm circle for drilling the holes for the cables. A centre hole should be drilled for the fixing bolts and rubber washers.

The capacitor should be mounted and some suitable terminals used. PL-259 and SO-239 plugs and sockets were used in the example, but ordinary spade terminals would be easier and just as good. The capacitor is wired to the connectors and loop using *only the fixed vanes*, the rotor being left unconnected or 'floating'. This halves the capacitance but doubles the working voltage. Further, it has the advantage of there being no moving contacts involved in the RF path. A well-insulated knob, or plastic shaft coupler and knob should be used for tuning.

When the whole framework is assembled, the loop can be wired. Because the conductor is formed in a cage, the cables forming this cage will be of different lengths. Each wire must be threaded through the holes in the plastic spacers *on the frame*. The 12 wires can be cut roughly to length with plenty to spare for termination (coloured wires are a great help here) or from a reel of cable. But, in either case, each wire must be run separately, like stringing a musical instrument. Start with the cables nearest the centre and work outward to the top. Terminate all 12 cables at one end first and solder them to a lug or terminal or PL-259 etc. Connect this end to the tuning box. The other end is more difficult, because the wires need to be reasonably tensioned to form the cage. If they are a little slack, the rubber washers holding the spacers to the spokes provide some adjustment for this purpose. Once the loop element is finished, don't forget to mark which is the top of the cage on the plastic spacers!

Great care must be taken with *all* soldered connections to minimise any DC resistance.

The coupling link is made with RG-213. Many trials were

Fig 1. The magnetic loop – for dimensions, see text.

THE G3HBN PORTABLE MAGNETIC LOOP

conducted to optimise the coupling but, with the 'fatter' conductor for the loop element, it was found that the coupling link also needed to be 'fatter'. To form the link, strip about 4cm off the cover, expose the inner conductor for 2.5cm and solder the braiding to the inner. From the tip of the join, measure 60cm of cable. Strip about 3cm of cover to expose the braiding and solder the shorted end to the exposed braiding. This will form the coupling link of about 19 to 20cm diameter. At the centre of the link, cut the braiding for about 2.5cm to expose the inner core as illustrated. Bend the remaining coax from the join to run vertically through the centre of the link and join it to the desired length of RG-58U. The mountings for the link are made with two Terry clips bolted back-to-back which clip neatly over the centre spoke and the RG-213.

Detail of the loop centre and link.

OPERATION

The first step in tuning the loop is to peak it for maximum aerial noise and/or signal strength on a receiver. If an aerial analyser is available, a quick check can be made for all the bands to observe the VSWR. A 1:1 VSWR should be obtainable on all bands from 7 to 29MHz. If 1:1 cannot be achieved, the aerial should be rotated, observing the VSWR at the same time. Adjacent objects in a room can unbalance the loop and, under these circumstances, it might not be possible to obtain the necessary VSWR. A VSWR/power meter is an asset, but the loop can be tuned with a simple field strength meter, tuning for maximum signal. Remember that maximum field strength radiation will be in the plane of the loop. On receive, this will be quite marked and a null will be obtained when the aerial is broadside to the signal. However, this does not always seem to be the case. Often very strong signals are received broadside to the loop and the same signal path seems to be effective for both transmission and reception. This may be due to building reflections, another illustration of where the loop needs to be easily and quickly rotated. This is particularly useful when operating low power (QRP).

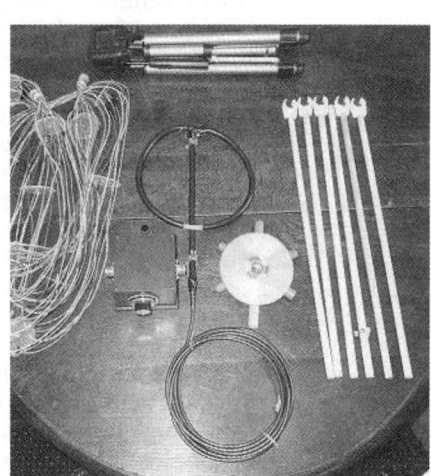

Ready for packing.

At the lower frequencies, the tuning of the loop is very sharp and it is essential to have a slow-motion drive fitted to the tuning capacitor. On the higher bands, the tuning is not quite so sharp, but it is still beneficial to tune the loop 'on the nose'. Hand-capacity has not proved to be a problem with this design. On the lower frequencies, the tuning capacitor is large and therefore any hand-capacity is insignificant. On the higher frequency bands, where hand-capacity is noticeable, it is still not too serious a problem because, at those frequencies, the usable bandwidth is much greater. There is, of course, absolutely no requirement for an ATU, if one is

WEEKEND PROJECTS

fitted to the equipment it should be either de-selected or tuned to a 50Ω load before using the loop.

The tuning box.

This model is not built for high power, but it will comfortably handle 20 – 25W. Normally, any indoor aerial used in a built-up area should not really be used for high power, since the problems of RFI (Radio Frequency Interference) can become critical. The loop should be placed as far away from the operator as is practicable.

RESULTS

Tests were carried out in CW (Morse) from the location pictured, in London, and from a seaside cottage in Folkestone, with power levels of 5W and 20W. The majority of contacts made were at the 5W level. The overall feel of the aerial was quite amazing, with signals over two S-units stronger than with the original 80cm loop, particularly on the lower-frequency bands. Comparison tests were made between the portable loop and the octagonal loop of 1m diameter on the roof. The roof aerial was, in the main, about 1 S-unit better, but the reports were that QSB (fading) was more prevalent with the indoor loop. Signal reports received varied with conditions, but reports of RST 579 and 589 were not unusual. Throughout the three-month trial period, frequent contacts were made with most European countries, Asia and North America. With 20W, the reply rate from stations called was between 70 and 80%; with 5W it was about 60 to 70% This is about the norm for QRP operation. Calling CQ was not very profitable and seldom is with QRP. Many two-way QRP contacts were also made, one notably on 30m with GM3OXX (1W RST 579), G3HBN being in Folkestone (5W RST 599), the loop being at ground level in the sitting room, about 20m above sea level. This was a very long ragchew. Conditions throughout the test period have been at an all-time low and extremely difficult for making reliable evaluations. However, several DX stations were worked with QRP and that in itself was most gratifying. No tests were made for RF feedback when using a microphone.

CONCLUSIONS

The object of improving the original loop has been achieved with greater success than expected. It seems the application of a multi-cable radiating element for the loop has brought with it more benefits than originally anticipated. The increased bandwidth is far greater than expectations and the improved overall performance in the liveliness of the aerial was a pleasant surprise. The next move is to replace the octagonal loop on the roof with a weather-proofed multi-conductor version. The results obtained here also open the door for further development in the general approach to the magnetic loop as an aerial in its own right, not necessarily to be compared with other aerial types. It is a radiator that has many characteristics that would seem not yet to have been fully exploited.

A UHF CORNER REFLECTOR AERIAL

The corner reflector is a well-known design and is capable of good performance on the VHF and UHF bands. At UHF, the practical implementation of the corner reflector is an ideal constructional project.

SOME DETAILS
Expressed quite simply, the aerial consists of a $\lambda/2$ dipole (where λ is the standard symbol for wavelength, making a '$\lambda/2$ dipole' a half-wave dipole.

Nothing new in that, you might say. However, the interesting feature is the reflector, which is not the usual single element, but a 90° metal 'corner', acting rather like a parabolic dish as used for satellite signal reception. The wind resistance of this type of reflector makes it impractical so, to reduce the 'windage' quite significantly, we make the 'corner' from closely-spaced rods, as illustrated in Fig 1.

The reflector consists of a number of 0.6λ rods, spaced from each other by 0.1λ. The aerial frame can be made of metal or wood, but wood is easier to work with, and mounting the elements to the frame is simpler. The prototype was made with wood of 20mm by 15mm cross-section, as Fig 2 shows. The wood was varnished for protection.

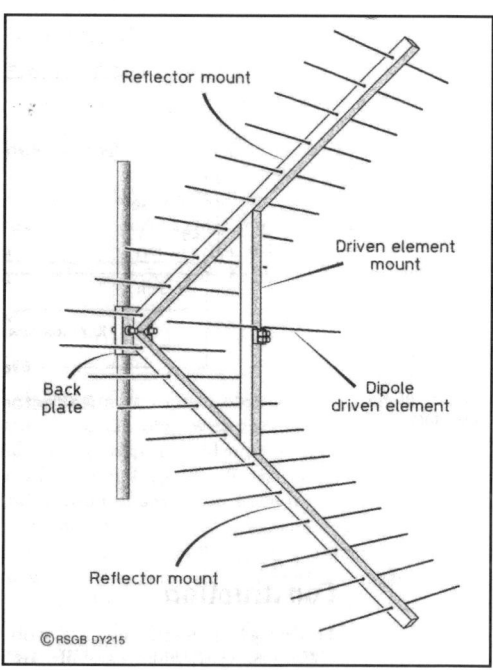

Fig 1. 70cm corner reflector aerial.

The elements were made from 1.5mm diameter copper wire, because a large reel of the wire happened to be available. The wire diameter is not critical; tubing could be used just as successfully. 14SWG hard-drawn copper aerial wire would be even better than that used in the prototype.

CONSTRUCTION
This project is just as much a woodworking project as a radio project! Follow the instructions carefully, and you should have little trouble.

- Cut the booms for the reflectors, as shown in Fig 2. A mitre block is invaluable here for producing the 45° corners.

WEEKEND PROJECTS

- Using the dimensions given on the diagram, mark the hole positions for the reflector elements, and then drill holes of a size which holds the elements firmly.
- Cut the driven element boom according to the diagram, and mark the point midway along the longer side, which will assist you later in positioning the driven element.
- Cut the back plate to size (about 120mm by 80mm). You may need to alter this size depending on the size of the U-bolt you will be using to clamp the aerial to the mast.
- If you want to be extra cautious in your construction, use the belt-and-braces approach, commonly known as 'screw-and-glue' to fix the booms to each other and to the back plate.
- Fix the reflector booms to the back plate first, then slide in the driven-element boom until it will go no further, then apply the wood glue and screw the two ends tightly to the reflector booms. Leave for the period prescribed by the glue manufacturers for the glue to harden.
- Varnish the whole structure.
- Cut the driven element to the correct size p1us a couple of centimetres (the reason for this will be evident in the 'Testing' section), and fix it to the centre of its boom (at the position you marked earlier) with a 'chocolate block' connector to which the coaxial feeder cable will eventually be connected.
- Cut and fix the reflector elements in place. If you find that these are a loose fit in the holes then, for each element, drill a pilot hole through the boom to intersect the hole for the element. File off the point of a woodscrew, and screw it gently into the pilot hole until it meets the element and grips it in place. You will now see why the point was filed off! Alternatively, you can glue the elements in place.

TESTING
Place the aerial on a mast, clear of obstructions. Connect it to a transceiver with a length of coaxial cable, with a standing-wave-ratio (SWR) meter in circuit. Find a clear frequency, identify your transmission and ask if the frequency really is clear. If so, key the transmitter again and note the SWR.

Do not stand in front of any aerial when it is radiating! The length of the driven element must be adjusted to obtain an SWR of less than two. If you have to shorten the dipole, bend the ends over rather than cut them off. That way, if you go too far, you can lengthen them again! The dipole was initially cut too long intentionally, to allow for adjustment here. Bending the ends over also reduces the risk of physical damage to clothing, skin and eyes. You may like to consider applying the same technique to the reflector elements for that reason alone.

MOVING ON
Once you have warmed to the idea of the corner reflector as an aerial, you might like to ring the changes regarding the reflector. How about

A UHF CORNER REFLECTOR AERIAL

replacing the 20 reflector elements with a wire mesh, such as garden centres sell as 'chicken wire'? Choose the finest mesh if there is a choice. Some extra support may be needed around the edges of the mesh, but you could go on to make a comparison of aerial gain between the two types, using a UHF field strength meter.

MATERIALS

Stiff wire or thin-walled tubing for dipole and reflector.
Frame – wood, 15mm by 20mm cross-section, lengths given in text.
Back plate – stout plywood, dimensions given in text.
U-bolt to suit mast.
Wood screws.
Wood glue.
50Ω coaxial cable for feeder.
2-terminal 'chocolate block' for dipole connection to feeder.
Varnish.

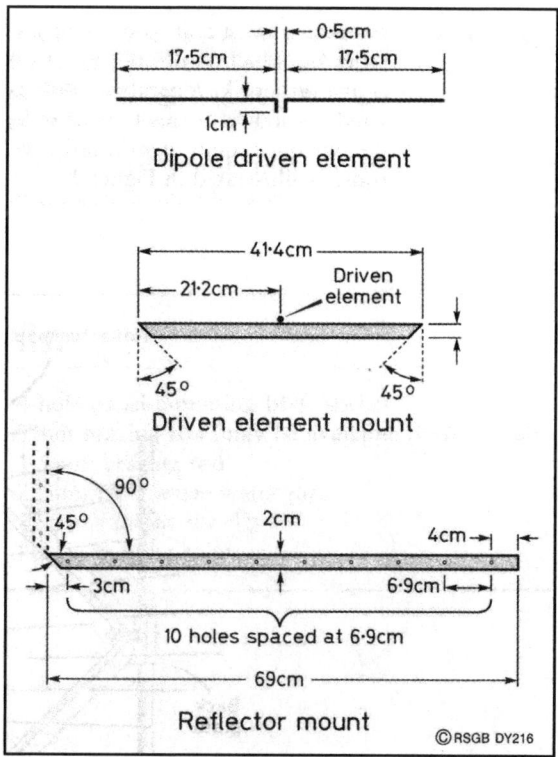

Fig 2. Driven element and boom dimensions for driven element and reflectors.

A TUBE YAGI FOR PORTABLE WORK ON 144MHz

This cheap, compact and simple-to-construct five-element Yagi is ideal for low-power backpack portable operation. The antenna elements pack up by sliding into the boom tube for easy storage. The antenna is light-weight and quick and easy to assemble in the field.

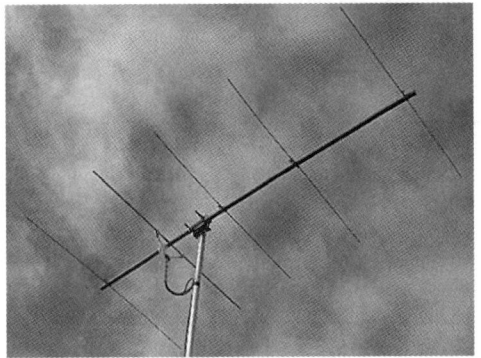

Photo 1. The 5-element Yagi unpacked from the tube and set-up on the portable mast.

A 177cm length of 19mm diameter plastic waste water pipe was used as the boom for the tube-Yagi. This needs to be carefully marked and drilled to make sure the elements will be parallel when they are inserted.

Start off by drawing a straight line down the tube. There might be a seam on the tube that you can use, otherwise try laying it on a flat surface and run a marker pen alongside using the surface as a guide. Next, wrap a piece of paper around the tube and use the straight edge as a guide to mark an accurate circle around the tube where the elements will go. Then cut the paper to the length of the circumference and find the halfway mark. By aligning one end of the paper with the line, the middle point will act as a marker for the holes on the other side of the boom. Don't try to drill through the whole boom in one go. Instead, drill the two holes separately. In this way you can correctly mark out, and then drill, the 10 holes on the boom that will take the five elements. You might need to open up the holes a tiny amount, but try to make the fit tight enough so that the elements slide in with a little friction.

Photo 2. Detail showing the last director with its labelling, the position of the stop and the grommet used to secure the element.

The reflector and director elements are made from 3mm aluminium rod (welding rod or similar). An off-centre blob of epoxy resin (or small circular push-on metal clip) acts as a stop for the rods so that they will be correctly positioned when inserted. A label can be added (to tag each element clearly for easy identification in the field) to this side of the element so that you will never try to put the element in the wrong way. Once the rods have been fitted into the boom, a small rubber grommet is pushed on from the other side to fix the elements in place (see the photograph).

A TUBE YAGI FOR PORTABLE WORK ON 144MHz

Tube of 6mm diameter was used for the dipole element, and a few turns of tape were used on one side to locate this.

The antenna is matched to the coaxial feeder using delta arms and a 4:1 coaxial balun to the dipole. The balun was constructed in the standard way (see Fig 1) using a loop of about 71cm. All the screens are soldered together at one point. The balun joints were covered in rubber solution glue to waterproof them and wired to solder tags to attach to the 15cm delta arms (which should be made of the same thickness tube as the dipole). The balun and delta arms were fixed to a sheet of plastic to provide support, as shown in the photograph. The other ends of the delta arms were attached 11.5cm either side of the dipole centre using 3M bolts and butterfly nuts. (For a really quick and easy connection, I have successfully used large crocodile clips instead of bolts and butterfly nuts (see photograph). However, this is only really suitable for low powers).

To pack up the antenna, the elements of the Yagi slide into the plastic boom. Rubber bungs stop the ends – you now have a Yagi-in-a-tube. I also found that it was possible to slide the tube-Yagi within the centre of my portable telescopic mast so that only one item then needed to be carried besides the backpack.

SETTING UP AND ADJUSTING THE SWR

On unpacking, the elements are fixed to their correct places on the boom and the rubber grommets used to secure the parasitic elements in place. The delta arms attach via the bolts and butterfly nuts either side of the dipole centre. When using the beam for horizontal polarisation, let the balun hang down away from the plane of the Yagi and tape the cable to the mast as shown in the photograph.

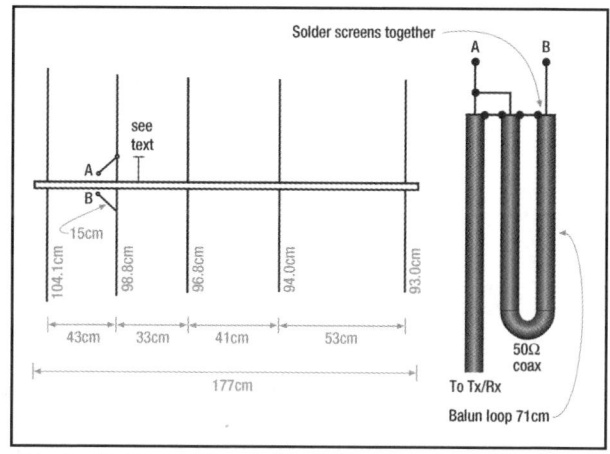

Fig 1. Schematic of the five-element Yagi and balun (dimensions taken from the ARRL article [1]).

Photo 3. View of the balun and delta arms. The view shows the croc-clip connections around the dipole centre used for initial testing, ideally these should be replaced with bolts and butterfly nuts.

WEEKEND PROJECTS

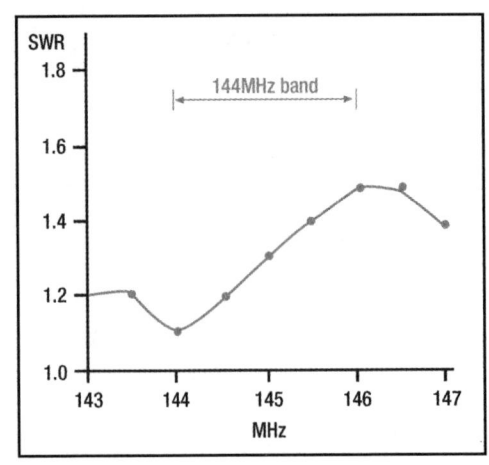

Fig 2. Typical SWR measurements for the tube-Yagi (horizontally mounted).

Adjustment of the SWR is possible by changing the lengths of the delta arms and where they attach either side of the dipole centre. I found that a good SWR was achieved over the whole of the band when the 15cm delta arms were attached 11.5cm either side of the dipole centre but some experimentation may be needed here depending what materials you decide to use. Shown in Fig 2 are typical measurements for the SWR of the beam when mounted horizontally. I used about 7m of RG-58A/U coax for this beam; lower-loss cable will probably change the curve slightly. The Yagi delta match and the balun are both resonant devices, so two peaks in the SWR can occur if the two resonances don't coincide exactly. A good SWR over a part of the band is easy and, with experimentation, an SWR of less than 1.6 over the whole 144MHz band is possible. For vertical polarisation, it is best to use a non-conducting mast. There is no natural position for the delta match / balun in the vertical position. I tried them fixed along the boom (toward the first director) and also perpendicular to the beam and each gave slightly different SWR results, but I found the whole assembly still gave very usable SWR results over the whole band, even with a metal mast.

ANTENNA GAIN

The antenna gain of the Yagi was determined in a very simple manner. The GB3VHF beacon in Kent (JO01DH), roughly 60km distant from the test site (IO90WU) was used as a reference signal and received firstly on a dipole and then secondly on the five-element Yagi (at the same height and with similar coax feeders).

A VHF attenuator (0 – 40dB in 1dB steps) was fitted between each antenna in turn and the receiver. The S-meter was used to measure the signal at the dipole. When the Yagi was measured, the attenuator was adjusted until the S-meter read the same as that obtained using the dipole. The change in attenuation thus gave a guide to the Yagi gain compared to the dipole. These experiments gave a gain of 7 – 8dB with a front-to-back ratio of 10 – 11dB.

Note: I used about 7m of RG-58A/U coax for convenience, as it is much easier to pack away. However, less loss would be obtained using the more bulky RG-213, but it is a heavier load for the antenna to support and so some support tape for the cable needs to be provided.

FINAL COMMENTS

A standard TV / FM aerial mast clamp was used to attach the Yagi to the mast. It is worthwhile replacing the nuts on these clamps for (butterfly) wing nuts; using them is quicker and cuts down on the number of tools you need to carry when out-and-about on the hills.

A TUBE YAGI FOR PORTABLE WORK ON 144MHz

The antenna is based on an old ARRL design [1] that I have used with excellent results over many years. The antenna seems to have a good combination of gain and bandwidth for its size. I have used the antenna singly, in twos and even in fours using a light weight open-wire phasing harness and universal stub, as described in the original article. Because the packed-up tube-Yagi is so compact and lightweight, it opens up the exciting possibility of being able to get further gain by using two (or perhaps even four!) of these antennas for backpack portable work.

PARTS LIST
1-off   plastic µPVC waste water pipe (eg 'Osma 5') 19mm.
1-off   ~1m, 6mm diameter aluminium tube (for dipole, see table).
4-off   ~1.1m, 3mm diameter aluminium rod (for passive elements, see table).
2-off   rubber bungs (to fit the ends of the boom).
1-off   TV / FM antenna clamp & wing nuts.
        Labels for element identification,
4-off   small rubber grommets & four cable ties.
2-off   solder tags and nuts & bolts.
2-off   3M bolts and butterfly nuts (or car battery size croc-clips, see text).
        50Ω coax (RG-58A/U convenient for portable use if short <10m) and RF connector for radio.

REFERENCE
[1]   *The Radio Amateur's Handbook*, (ARRL), 1972, Chapter 22 – 'VHF and UHF Antennas'.

For the **best** selection of **Amateur Radio books**

Only £29.99 plus p&p

RSGB Radio Communication Handbook

9th Edition, edited by Mike Dennison, G3XDV & John Fielding, ZS5JF

The Radio Communication Handbook is bigger and better than ever before!

Extensively revised, the 9th edition of the *RSGB Radio Communication Handbook* has once again had a major update. Every chapter has been enhanced and several have been re-written by acknowledged experts. Just about everyone will find items of much value in this great book Chapters vary from the essentials right through to detailed ones on specialist topics. The book also contains for the first time a 52 page chapter on Software Defined Radio from the RSGB columnists Steve Ireland, VK6VZ and Phil Harman, VK6APH. The "Low Frequencies" chapter now includes projects for the experimental frequencies near 500kHz. The chapter covering "Computers in the Shack" has been completely revised to include DSP, Bluetooth, USB and PIC programming. "Construction and workshop practice" has been re-written, whilst "Data Communications", now includes several new HF data modes. Appendices contain all the useful reference data and artwork for printed circuit boards. With 26 chapters spread over 800 pages this book is packed with far more ever than can be detailed here.

Bonus CD: You can also search every page of the *RSGB Radio Communication Handbook* at the touch of a button with the enclosed CD. It not only contains a searchable and printable PDF version of the book but there is much more. There is a bonus chapter, an SDR video, extra detail and a host of amateur radio software.

- Biggest ever Handbook: over 800 A4 pages
- 26 chapters and two appendices
- Half a million words!
- Over 1500 illustrations
- LF including 500kHz
- Software Defined Radio
- Projects for 136kHz to 76GHz
- Bonus CD

Covering the entire spectrum from the basics through to advanced projects, and including many classic circuits, the *Radio Communication Handbook* makes an essential shack accessory. If you only ever buy one book on amateur radio, this should be the one!

Size 297x210mm, 800 pages   ISBN 9781-9050-8633-7

Radio Society of Great Britain
Lambda House, Cranborne Road, Potters Bar, Herts, EN6 3JE Tel: 0870 904 7373 Fax: 0870 904 737

www.rsgbshop.org

RSGB SHOP

GENERAL

Using 10GHz 34
A simple converter for the 10m satellite band 39
An 80m transceiver 47

USING 10GHz

Mention microwaves and many beginners will say "It's too complicated, too costly and too difficult for me". Well, present-day, state-of-the-art, narrow-band equipment designs almost certainly are too difficult and possibly too costly for beginners. Many newcomers do, indeed, appear to be put off by this high technology.

So where do beginners start? This article may help to revive interest in the simple, easy, and inexpensive wideband FM (WBFM) designs that most of today's advanced operators used before the latest state-of-the-art designs became available for experienced home constructors.

The ideas are not new, so don't expect a 'blow-by-blow' description. Full details are available in the references quoted. The intention is to encourage you to get hold of this information, have a go, produce working equipment, get the feel for microwave operating and then, perhaps, to have a go at some of the more advanced microwave designs as your skills and knowledge improve with practice. Believe me, you *don't* have to be a skilled constructor or operator to take the first steps.

WHAT'S NEEDED?
You will need the following major 'ingredients' to construct a 10GHz WBFM transceiver:

- An aerial (see, for example, the horn aerial in reference [2], pp34 – 41)
- A mast (for fixed station working) or tripod (for portable working see reference [2], pp107 – 108)
- An in-line Gunn oscillator/mixer Doppler intruder alarm module (see reference [1], pp98 – 102)
- A Gunn oscillator power supply/modulator module (see reference [1], pp94 – 97)
- A low-noise, wide-band IF preamplifier module (see reference [2], pp122 – 124 or pp125 – 127)
- An FM receiver to tune the chosen IF (see, for example, reference [2], pp126 – 136 – there are many alternatives, see later)
- Various screened (metal) boxes, switches, plugs and sockets, controls and control knobs, cables, wire and general hardware – see later

USING 10GHz

- A 12V power supply at 1.5A maximum (including a scanner receiver), battery for portable use, mains for fixed station use.

So, how does it all work?

THE RECEIVER

A basic superheterodyne receiver for *any* frequency consists of two main parts, as shown in Fig 1. The first is a *front-end converter* consisting of a local oscillator (LO) and a mixer to produce a lower intermediate frequency (IF), the sum or difference of the LO frequency and the received frequency. The reason for converting the high-frequency received signal down to a lower frequency is because it is easier to amplify, filter and demodulate. The second part is a *back-end*, consisting of IF amplifiers, filters, often a second oscillator/mixer (to convert signals to a second, even lower IF), a demodulator and an output stage.

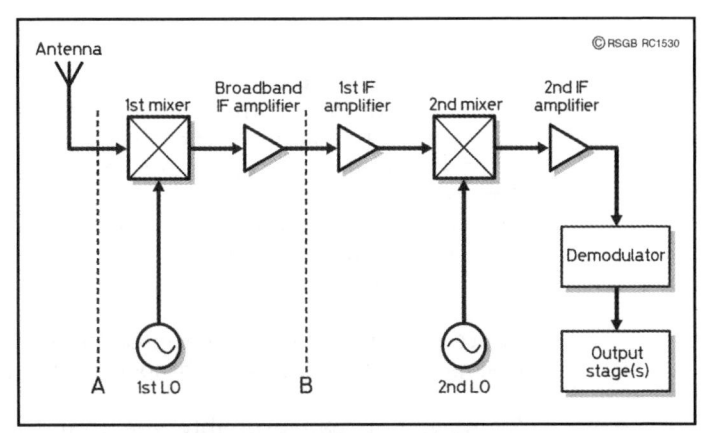

Fig 1. Block diagram of a basic double-conversion superheterodyne receiver. The front-end converter is between A and B, and the back-end is all to the right of B.

A 10GHz receiver is no different. It would be very difficult for a beginner to make a microwave oscillator and mixer. Fortunately, there are ready-made oscillator / mixer units available in the form of surplus 'in-line' Doppler intruder alarm units. These are obtainable from most amateur radio rallies for £5 or £10. They consist of a Gunn device oscillator and a diode mixer mounted in-line inside a short length of (usually) waveguide 16 (WG16) which forms *cavities* (tuned circuits). WG16 is the standard waveguide size for 10GHz (3cm).

An aerial is connected to the open end of the waveguide and the received signal collected by the aerial travels down the waveguide to the mixer diode. LO input to the mixer is via a hole or slot in the end wall of the LO cavity. The IF output is taken from the mixer diode terminal on the Doppler module and connected to a suitable receiver back-end.

The back-end of the receiver can be almost anything you like, provided that it will tune to an IF somewhere between 10.7MHz (minimum) and, say, 50 or 60MHz, and has the necessary amplifiers, filters, demodulator and output stages. IFs below 10.7MHz can't be used because of LO noise, while IF signals above about 60MHz may be bypassed by the mixer diode decoupling capacitor built into the Doppler module. At one time a favourite choice was an FM broadcast band receiver covering

35

88 to 108MHz: these days there are so many FM stations in that band that it is almost impossible to avoid IF breakthrough.

Several choices can be made: a complete 10.7MHz back-end (see reference [2], pp126– 136), a simple 50MHz receiver (see reference [2], pp74 – 86), or a scanner tuned to whatever IF seems to work best. The choice is yours.

THE TRANSMITTER
The power output of the Doppler module Gunn oscillator is usually in the range 10 to 20mW. The mixer diode uses only a small proportion of this, leaving the remainder to escape from the open end of the Doppler module which of course is connected to the aerial. Since both the receiver input and the transmitter output (the open end of the Doppler module) are connected to the aerial, we don't need to switch the aerial between transmit and receive.

All we need in order to use the Gunn oscillator as a transmitter is some means of modulating it with speech or an audio tone.

By now, you can probably see that the heart of a simple 10GHz transceiver is the Gunn oscillator used both for receive and transmit. The Gunn oscillator needs a stabilised, variable, low-voltage bias power supply (typically 6 to 9V at about 150mA) to operate. If you want to know exactly how a Gunn device works, refer to reference [3]. Two factors determine the exact Gunn oscillator frequency – the size of the cavity (tuned circuit) in which it is mounted and the bias voltage which is applied.

The cavity can be tuned through several hundred megahertz by means of a metal screw in the cavity wall near to the Gunn device. This is a useful way of coarse tuning; inserting the screw into the cavity lowers the oscillator frequency, while withdrawing it raises the frequency.

Varying the Gunn bias voltage also changes the oscillator frequency. This is an effect known as *frequency pushing*. Lowering the voltage decreases the frequency, while raising it increases the frequency. The effective tuning range is limited to a few tens of megahertz, so that this is a very convenient way of 'fine tuning' the oscillator. A few millivolts of audio or tone modulation can be added to the bias voltage to produce high-quality FM modulation. The Gunn bias (power) supply therefore needs to be variable but, once set to a particular voltage, must be stable and free from noise. An unstable voltage or a noisy supply will result in an unstable and noisy oscillator! A very reliable Gunn supply with modulation facilities was described in reference [1], pp94 – 97. Use a 10-turn potentiometer and turns-counting dial for the 'fine-tune' control – this gives good bandspread and frequency-setting accuracy and is well worth the extra cost. The turns-counting dial can be calibrated in frequency to make frequency setting easier.

OTHER CONSIDERATIONS
Two other things need to be considered. Problem number one: coaxial feeder can't be used between the aerial and the Doppler module

USING 10GHz

because the feeder losses at 10GHz would be astronomical! Problem number two: the IF output level of the mixer is very low so, again, losses in any connecting cable are not acceptable. What's to be done?

Both problems can be quite easily resolved by building the transceiver in two units. Mount the Doppler module and wide-band amplifier in a waterproof, screened (metal) box, attach the aerial directly to the same box and mount the whole of this 'microwave head' at the top of a mast for fixed station use – see reference [1], p94, or reference [2], p126, or on a tripod for portable use.

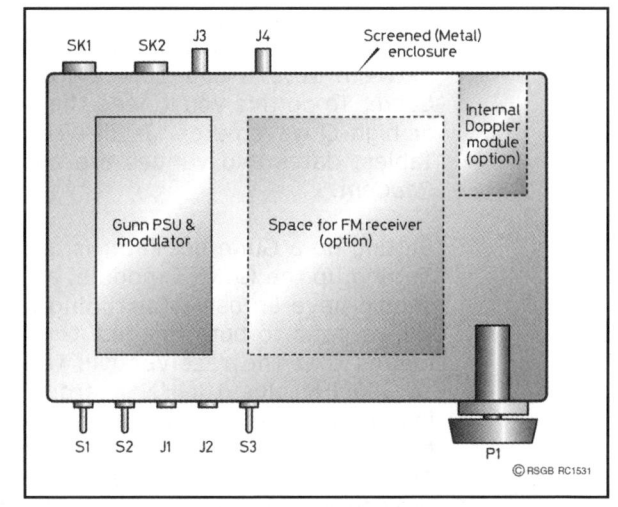

Fig 2. A suggested layout for a base unit, extended to include a switchable internal Gunn.

Build the Gunn power supply / modulator and all the controls (including the receiver back-end) into a base unit which can be used in the shack for fixed-station use or in the car for portable use.

Of course, if you want to work both fixed and portable, you will want to leave the mast-mounted microwave head at home and use a second, identical head for portable use. Build two masthead units or build another Doppler module into the base unit and use simple switching to change from the 'local' to the 'remote' microwave module, as suggested in Fig 2. S1 is the power on / off switch, S2 selects speech / tone, J1 is the mic jack, J2 the key jack, J3 coax to external receiver, J4 coax to mast-head unit, SK1 is for 12V in, SK2 for 12V to external receiver, P1 is the Gunn 'fine tune' control – 10-turn potentiometer / turns counter. The additional control S3 switches internal / external Gunn.

Built like this, one ordinary coaxial cable can now be used to connect the base unit to the remote unit, sending the modulated Gunn bias up to the remote microwave head and the amplified (broad-band) IF down from the remote microwave head to the base unit.

Full duplex operation was mentioned (see reference [2], p125). Unless the two stations share an identical IF, this will not be possible and it is probably the exception rather than the rule. A simple addition, shown in Fig 3, allows the G4KNZ Gunn PSU / modulator to be set for fixed-frequency transmit, while allowing receiver tuning over the whole range of frequency pushing. When everything appears to be working correctly, the last things to be done are to retune the

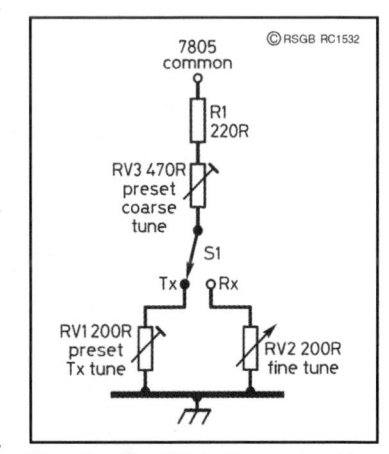

Fig 3. Modifications to the G4KNZ Gunn PSU / modulator to give a fixed transmit frequency and a tunable receiver.

37

WEEKEND PROJECTS

Doppler Gunn oscillator from its usual ISM (Industrial, Scientific and Medical) frequency of 10,687MHz into the amateur band, somewhere between 10,370 and 10,400MHz, which is the current UK WBFM sub-band. To do this you'll need the help of someone with a 10GHz counter or high-Q wavemeter. If all else fails, try one of the Microwave Round Tables; dates and venues are announced in the 'Microwave' column in *RadCom*.

Setting up a Gunn oscillator is described in reference [2], pp134 – 135. Setting up the G4KNZ module, as modified in Fig 2, is very similar. You should have an oscillator tuning range of about ±10MHz. The receiver will respond to both images, that is, $f_{osc} \pm F_{IF}$. If the IF is 10MHz, this means that the receiver will tune from 10,370MHz minus 10MHz to 10,390MHz plus 10MHz, ie from 10,360 to 10,400MHz. Using either image means that you can tune both the narrowband segment (10,368 to 10,370MHz) and most of the wide-band segment (10,370 to 10,410MHz). You might be lucky enough to hear a beacon which will provide a very accurate frequency marker to check your calibration.

CONCLUSION

It is possible to construct an entirely practical 10GHz WBFM transceiver using ready-made 10GHz Doppler oscillator / mixer modules, with some kind of FM receiver back-end. The overall cost of the whole project will depend on how much you like to experiment, how elaborate a transceiver you want to make, how carefully you shop around for parts and, to some extent, on how big a junk box you have! The point is that you do not have to be a skilled constructor or operator to get results.

Don't expect to work the world with simple, low-power 10GHz WBFM, but do expect to have a great deal of fun and gain useful experience and skills in trying to work your pals, either from home or portable (over much longer distances) using the 10mW or so transmitter power that typical, simple WBFM equipment produces.

Since there's plenty of bandwidth available, Doppler modules can be used for amateur fast-scan television (ATV) or very high-speed packet links – but that's another story. Think about it – microwaves have a lot to offer.

REFERENCES
[1] *Practical Transmitters for Novices*, GW4HWR, RSGB.
[2] *Practical Receivers for Beginners*, GW4HWR, RSGB.
[3] *Microwave Handbook*, Vols 1, 2 and 3, ed G3PFR, RSGB.

A SIMPLE CONVERTER FOR THE 10m SATELLITE BAND

Many amateur radio satellites pass by every day. They re-transmit signals from amateur radio ground stations and each satellite generates at least one beacon signal. The signals from the amateur radio satellites in the 29.3 to 29.5MHz segment of the 10m band are usually CW or SSB. They are converted by this simple unit to the 80m band for reception with simple receivers. The converter will work with any receiver that can tune from 3.5 to 3.7MHz and which can receive CW and SSB signals.

WHAT'S UP THERE?

Several Russian satellites make regular and frequent daily passes, some 19 to 22 passes each day in polar or low-earth orbit (LEO) at altitudes around 1000 and 2000km. Two satellites may pass by at the same time! The beacon signals from these satellites are consistent and easy to find.

This converter is sensitive and a preamplifier is not necessary. Satellite beacon signals can be heard in the noise for a short period as a satellite approaches from over the horizon at the predicted time. The signals are well above the noise for most of a pass, which can be from only a few minutes to 30 minutes, until the satellite disappears back into the noise and beyond the horizon, to reappear again some 100 or so minutes later.

WHAT DO THEY DO?

The satellites transmit *downlink* signals, re-transmitted from amateur stations on the ground, and operating in *Mode A* (2m uplink to the satellite, 10m downlink from it).

This simple receiving set-up is an introduction to satellite communications. It can lead to the study of satellite orbital and celestial mechanics and to the decoding of satellite telemetry signals. Information about the status of the satellite and its hardware is sent in the beacon transmission and is available to any listener. It is easily logged and decoded. The trends of several of the satellite's parameters can be followed.

There are many satellite passes, so the chance of hearing a satellite by just listening at random times is reasonably high. Computer programs for accurately predicting the times of satellite passes are freely available. More about those later.

WEEKEND PROJECTS

Listening to amateur radio satellites leads to many exciting self-training activities: building and operating the receiving equipment; listening to beacon transmissions from satellites; monitoring amateur stations working through the satellite; displaying the current position of the satellite on a map on a computer screen; predicting the satellite's orbit by computer; recording and decoding the telemetry information to follow changes in the spacecraft's status, and much more. All these activities develop from, and are supported by, the signals received by a simple receiver and its attached converter. These activities are the things that school science and technology projects are made of – investigatory work requiring discipline, organisation and record-keeping, essential to budding engineers and scientists.

THE CIRCUIT

This simple converter circuit, shown in Fig 1, is based around the popular NE602 integrated circuit. A 6V battery provides the energy supply.

There is only one control – a toggle switch. It acts as a band switch to switch from the 80m band to the 10m satellite band. Three poles of a miniature four-pole changeover toggle switch are used. One pole acts as the battery on/off switch (S1b) with two poles (S1a and S1c) switching the converter in and out of circuit. Remember to switch the unit off, when not in use, to conserve the batteries!

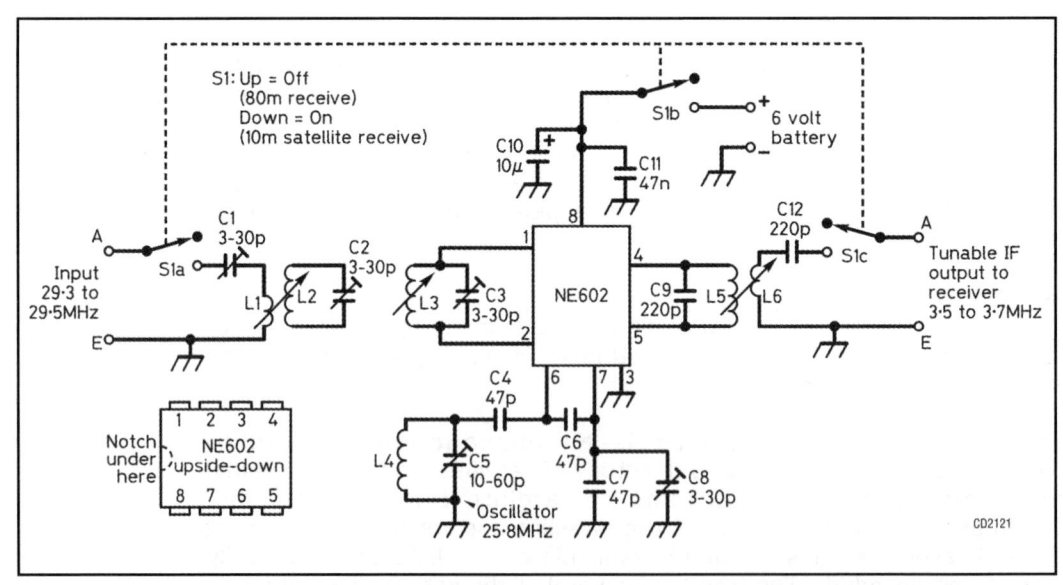

Fig 1. The circuit of the 10m converter.

THE OUTPUT CIRCUIT

Starting at the output, IC pins 4 and 5 feed a tuned circuit, broadly tuned to 3.5 to 3.7MHz. A link coil, L6, feeds the output signal through a short length of twin-lead or twisted pair to the receiver aerial and earth terminals. Capacitor C12 is for matching purposes. Coaxial cable is not necessary for the link to the receiver.

A SIMPLE CONVERTER FOR THE 10m SATELLITE BAND

L5 is 30 turns of 26-gauge enamelled wire on a 6mm slug-tuned former. The same wire gauge is used for all coils in this converter. The link coil L6 is 6 turns wound over the bottom of L3. An untuned toroid transformer is an alternative for this output arrangement. The tuned circuit was adopted to give some added IF selectivity in the expectation of the possible appearance of out-of-band unwanted signals. This has not been a problem.

THE OSCILLATOR

IC pins 6 and 7 connect to the oscillator section. This is a free-running LC circuit. The oscillator frequency is 25.8MHz but more about this selection later. Cheap quartz crystals are not readily available for this frequency, so a Colpitts LC oscillator has been used as shown in the diagram.

The coil L4 is 10 turns on a 6mm former. The slug has been removed from the former. For stability reasons, the coil and former are enclosed in a soft plastic foam jacket held in place by a metal saddle soldered to the PC board. An earthed metal shielding can to contain the coil completely could be used. The trimmer capacitor C8 is mounted so that it can be adjusted with a trimming tool from outside the converter unit through a hole in the box, should a slight frequency change be required.

The oscillator is surprisingly stable. During construction, soldering and mechanical work on the oscillator components puts stress on those components. After several days in a closed box, the oscillator frequency settles down and stabilises. It wanders very slowly in the range about 5kHz above and below 25.8MHz during each day, irrespective of whether it is left switched on, or only switched on when it is required. The unit should not be left in direct sunlight. No special selection of components has been made to improve the frequency stability.

The oscillator frequency varies by less than 100Hz during a long satellite pass. This has been found to be quite acceptable. The Doppler frequency shift observed on the incoming signals is very much more than this, so any steady long-term oscillator drift in such a simple unit is tolerable. The receiver tuning control is swung across the satellite downlink band during a pass looking for signals and continuous tuning adjustments are made to correct for Doppler shift, so the absolute frequency at any time is not of major importance. The initial reception of the satellite beacon signal gives an indication of any dial error for that particular pass.

THE INPUT CIRCUIT

The twin inductors L2 and L3 with their associated capacitors form the input tuned circuit, tuned to the 29.3 to 29.5MHz satellite band. This band-pass filter configuration was chosen to minimise strong local 80m signals. The circuit shown has made the occurrence of breakthrough rare.

There can still be a problem at times with exceptionally strong local 80m stations. Traps for these signals could be added to the converter but the added complexity is not warranted. Upper sideband is used on

WEEKEND PROJECTS

the satellite downlinks. This is different from the 80m band, where lower sideband is in general use, so it is easy to identify any LSB signal which may be heard as IF breakthrough. Switching to 80m reception will confirm if the signal is on the 80m band. The problem is insignificant.

The coils L2 and L3 are identical – each is 12 turns wound on a 6mm former. An input link L1 of three turns is wound over the bottom of L2. The coil formers for L2 and L3 are mounted side-by-side, in parallel, about 20mm apart, centre-to-centre.

CONSTRUCTION

The unit is built in a plastic box about 150mm wide by 55mm high and 90mm deep, available from many sources. A smaller box could be used. A completely sealed box is needed to keep air movements away from the oscillator components and to increase thermal delay in the interests of oscillator stability.

The IC is glued upside down to a piece of PC board and wiring is by the 'dead-bug' method. Tag strips support other components and the coil formers. The PC board is bolted to the lid of the box, which is then positioned upside down, so the lid becomes the base.

The two critical things about the wiring are to keep the signal input and output leads well separated (to minimise break-through of 80m signals), and to make sure that the leads to the oscillator components are rigid (to IC pins 6 and 7). Use the *outside* sections of the four-pole toggle switch for S1a and S1c to improve isolation of the input and output signal leads.

As with all battery-powered devices, be certain that the batteries are connected with the correct polarity!

THE FREQUENCIES INVOLVED

With the 25.8MHz oscillator in the converter and the receiver tuned to 3.5MHz, reception will be at 25.8 + 3.5 = 29.3MHz. Tuning 200kHz higher to 3.7MHz, it will receive at 29.5MHz. These two frequencies can be useful in calibrating the tuning scale.

Other oscillator frequencies could be chosen to give other tuning ranges on the 80m band and would be just as effective. If loud local 80m breakthrough is a persistent problem at the satellite frequencies of interest to you, the oscillator could be shifted 100kHz lower to 25.7MHz to make the satellite band read from 3.6 to 3.8MHz on the receiver dial.

SETTING-UP

Check the circuit and your wiring. Connect the converter to the receiver and switch on. Coil L5 is peaked for maximum noise at 3.6MHz – the middle of the reception range.

The oscillator trimmer capacitor C8 is set to a half-way position. The oscillator tuning capacitor C5 is then adjusted to set the oscillator to 25.8MHz.

A SIMPLE CONVERTER FOR THE 10m SATELLITE BAND

Listen at that frequency on a communications receiver with a digital readout and adjust C5 until you can hear the oscillator carrier in the centre of the pass-band. The trimmer capacitor C8 is used for fine adjustment through a hole in the box. Put a piece of tape over the hole to keep the box sealed. If you don't have access to a communications receiver, try a push-button FM broadcast receiver; even a push-button FM car radio is suitable. Hold the converter near the FM receiver aerial. Listen at 103.2MHz. This is the 4th harmonic of 25.8MHz. You should hear the FM noise decrease with a swish as the converter oscillator is adjusted. Set the converter oscillator to full quieting on the FM receiver; the correct position is easy to identify.

Due to slight variations in the values of the fixed capacitors used, it may be necessary to add or remove a turn to your coil to get the oscillator to the correct frequency.

Attach the aerial lead! The three input tuned circuit trimmer capacitors C1, C2 and C3 are all set to a halfway position. The receiver is set to 3.6MHz for this adjustment, with the incoming 'noise' peaked at 29.4MHz – the centre of the satellite band. The slugs in the formers L2 and L3 are adjusted for maximum received noise. The trimmers C2 and C3 are for fine adjustment – the noise peaks are sharp.

Be aware that there may be more turns on the coils L2 and L3 than necessary; it may be possible to set the slugs too far into those coils and obtain a noise peak at 25.8 - 3.6 = 22.2MHz.

The input capacitor, C1, is then slowly moved to its minimum capacitance position. The noise heard should decrease. Advancing the trimmer back towards maximum capacitance will increase the noise level. The noise will be found to increase and level off. Set the trimmer to the point where the noise appears to first reach this plateau. Too much capacitance will be an invitation for 80m breakthrough to be a nuisance. This adjustment will depend upon the characteristics of the connected aerial. The setting is not critical.

Now listen for a satellite beacon signal at the frequencies given below. When a satellite beacon signal has been identified, it can be used to adjust the input trimmers C2 and C3 even more accurately. Once correctly set up, there is little need for any further changes to any of these settings.

SATELLITE RECEPTION
The International Radio Regulations of the International Telecommunication Union show that the whole 10m band (28 to 29.7MHz) is available to the Amateur Satellite Service. Amateur radio band planning has assigned the band segment 29.3 to 29.5MHz to exclusive amateur satellite use. The frequencies for each of the RS series of satellites are given in the amateur radio literature. The details you need to get started with each satellite are:

RS-10: Beacon 29.357 or 29.403MHz; downlink 29.360 to 29.400MHz.
RS-12: Beacon 29.408 or 29.454MHz; downlink 29.410 to 29.450MHz.

WEEKEND PROJECTS

RS-15: Beacon 29.3525 or 29.3987MHz; downlink 29.354 to 29.394MHz.

The first beacon frequency listed for each satellite seems to be the one in general use.

These frequencies show our band of interest to be 100kHz wide and centred on 29.4MHz. So it is appropriate to use the RS-12 beacon at 29.408MHz to peak up the converter's input tuned circuits C2 and C3 finally for optimum performance.

The RS-16 satellite is now in orbit and adds six or seven more satellite passes each day for the keen listeners to the 10m satellite band. Its beacon and transponder output frequencies are given as being similar to RS-12.

UNDERSTANDING THE TELEMETRY

The beacon gives repeats of "CQ de..." and its RS number, followed by text and telemetry information, in Morse code. For example, the telemetry from RS-10 is a frame of 16 four-character words. By carefully noting and processing this detail, one can obtain an interesting appreciation of the goings-on inside the spacecraft. There are 16 monitoring points and 16 measurements are sent in this repeating cycle.

A book listed in the references below contains information about how to interpret the RS-10 telemetry information. RS-12 is said to be a copy of RS-10. The RS-15 telemetry information is different. It is given in a document downloadable from the AMSAT Internet website, and the details follow below.

Most satellite beacon transmissions on this band are in CW. The Morse code speed may be too fast for the novice. There is a simple way to overcome this. Use a two-speed tape-recorder. Receive a beacon transmission using a high-pitch CW tone and record it at the fast tape speed. Play the recording back later at a slower tape speed!

SUITABLE RECEIVING AERIALS

A simple length of wire is a basic aerial for HF reception. So the same random length of wire used for many experiments, about 20m long, strung out of the window, is used for reception of these satellite signals. It is probably too long for the purpose but it works! A half-wave dipole aerial cut to length for the 10m band, positioned up clear of obstacles and centre-fed with coaxial cable, or a vertical 27MHz CB aerial, may be better receiving aerials. Experiment! There are many good aerial textbooks that can assist.

AN AUDIO FILTER

An audio filter is a useful accessory for any receiver when seeking narrowband signals in noisy conditions. Weak signals from approaching satellites, rising up from the noise at a distance of more than 4000km, make a supplementary audio filter useful but not essential.

A SIMPLE CONVERTER FOR THE 10m SATELLITE BAND

At times it is possible to hear a 10m satellite beacon signal at very long distances away, well beyond the horizon. This is due to the characteristics of HF propagation. An audio filter is useful here to help to identify these signals in the noise.

WHEN TO LISTEN?

Three ways are suggested to find the times of arrival of the satellites. The first is to listen at random times over, say, a two-hour period. These satellites have an orbital period of some 105 to 128 minutes. With some 20 passes each day each of up to 30 minutes' duration, there is a chance with random listening that you will hear signals!

A second way is to ask a local amateur radio satellite enthusiast to provide you with a computer printout of predictions for the 10m band satellites for the next few days for your location. This will give you the times to listen and a lot of other details about the position where the satellite comes up over your horizon and where and when it disappears.

A third and preferred method is to set up a personal computer and a suitable computer program to produce this orbital data for yourself at the time you want it. The computer is a very useful tool to support many amateur radio activities. A computer satellite prediction program will show when a satellite of interest will be visible from a specified location. This may be by graphical means with a map, or by a table of times and directions which can be printed out. Such programs give the AOS (acquisition of signal), the time when a satellite clears the horizon to come within your line of sight, and the LOS (loss of signal), the time when the satellite dips below the horizon.

The time you first hear the beacon signal from an approaching satellite depends on several things, including hills and other obstacles that may obscure your view of the computed horizon. There are occasional satellite passes in which a satellite, at a range of more than 4000km, rises only a degree or so above the computed horizon and for only a few minutes. These distant passes may be obscured by high hills on the visual horizon in that direction. Reception of the satellite signals with any receiving equipment in those circumstances, if it is possible, will probably be weak and noisy. There are several good computer programs for satellite prediction available by download from the Internet. Some are free; others require a small donation. Start at the AMSAT website, **www.amsat.org**. AMSAT is a nonprofit-making organisation of volunteers promoting and supporting amateur radio satellite activity.

Some programs have excellent graphics but may require as a minimum a 386 or better computer. A good explanatory document accompanies some programs.

To remain current, all satellite computer programs need regular updating of the orbit information by entering new *Keplerian elements*. These are explained in simple terms in documents available for download from the AMSAT website.

WEEKEND PROJECTS

Regular Keplerian element update information is available from amateur radio packet bulletin boards and from various Internet sources. If you have electronic mail facilities, you can put your details forward at the AMSAT website to get on a free Internet mailing list to bring Kepler updates direct to your computer by e-mail message each week. This is an AMSAT service to promote interest in satellite activity. Most satellite programs have procedures to accept Kepler updates automatically from messages in a two-line format, so tedious typing and error-checking are not necessary.

AMSAT needs our financial support for the hardware in the sky and for the resources of technical information so readily provided. Details about the various AMSAT groups are on the website and it has many downloadable documents for the budding satellite enthusiast. One gives a very full description of the RS series of satellites.

With a computer program correctly installed for your location, and with the appropriate current Kepler elements, the position and ground track of several satellites, such as the Russian RS series, the Space Shuttle, the Hubble Space Telescope and the Gamma Ray Observatory, can be obtained. Many of these satellites can be observed visually with the naked eye in the dark of the evening as they pass by, weather and other conditions permitting. With an amateur radio licence and a VHF transceiver, you can talk to radio amateurs on the International Space Station!

MODIFICATIONS?
A dedicated receiver could be built for the 10m satellite band. This idea was considered and rejected in favour of the converter approach. The converter can be made fixed-tuned for the relatively narrow satellite segment and the receiver is still available for 80m reception.

IN CONCLUSION
Satellites are a very interesting aspect of amateur radio and it is very easy to get hooked. Just build a simple 80m receiver with a satellite converter. It is a great thrill to receive satellite signals using equipment that you yourself have built. This can lead to so many other things to investigate. Satellite pass times soon dictate your daily routine. A challenge indeed! Give it a go, and have fun – there is much experimenting to do!

RECOMMENDED READING
The ARRL Satellite Anthology, 3rd edn, ARRL, 1994, pp114–119 inclusive. This is the best of *QST* articles on amateur satellite operation and hardware. Telemetry decoding for RS-10 is on pages 115 and 116.

AN 80m TRANSCEIVER

This transceiver was developed as a project that would hopefully appeal to the relatively inexperienced constructor, although it is probably not ideal for the complete beginner. The ability to handle a soldering iron and identify components is necessary, as is the ability to read and work from a circuit diagram, but a lot of detail is included to make construction as straightforward as possible.

DESIGN PHILOSOPHY

The aim was to produce a basic transceiver with minimum component count that required little in the way of tools or test gear to get working, but which would give creditable performance.

A tall order, perhaps. The first step was to decide exactly what was required. The obvious choice was a single-band QRP rig. On the receiver side, I opted for direct conversion, as a superhet design would be far more complex. For the same reason, I decided on CW rather than SSB.

Now, if you are going to operate QRP CW for the first time, the ideal place to do it (in my opinion) is on 80m. On this band, a watt or two will give you QSOs all over the UK and well into Europe, a fact that I discovered around 30 years ago, before QRP operation really took off. I reduced the output of an old valve transmitter to about 3W, and to my surprise found that it seemed just as easy to make QSOs with this as it did with my 100W transmitter.

The completed transceiver (centre).

WEEKEND PROJECTS

There are many circuits for this kind of transceiver around. A lot of them seem to be severely limited by over-simplification. I consider that VFO (variable frequency oscillator) control is highly desirable, but that RIT (receiver incremental tuning) is absolutely essential. If you are using crystal-control for transmit and receive, you have to rely on the other station being slightly off your frequency in order to produce the necessary beat note, or be able to shift the frequency of your oscillator slightly. With RIT and a VFO, it is a simple matter to 'net' onto another signal, and then adjust your RIT for a comfortable note. Another highly desirable feature is a sidetone oscillator. This enables you to hear what you are sending, and does not add greatly to the overall complexity.

The method of construction is at least as important as the electronics in a practical project.

I opted to use double-sided copper-clad board to make a chassis / panel arrangement, with a rear apron and a screened enclosure for the VFO. Cutting and drilling are kept to a minimum, and the need to manufacture printed circuit boards avoided by mounting most components (with the exception of those in the VFO, which are mounted on a small piece of matrix board) in a 'dead bug' fashion. More on that later.

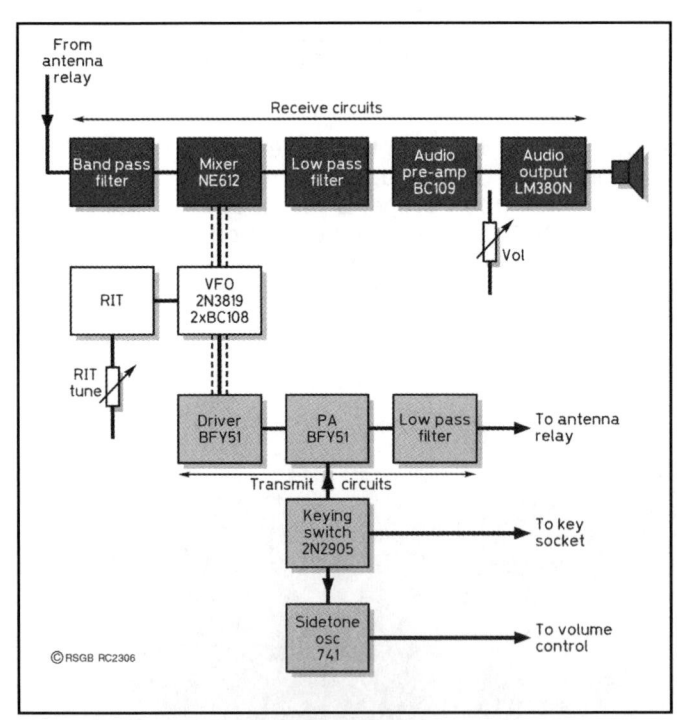

Fig 1. Block diagram of the 80m CW transceiver. Power supplies have been omitted for clarity.

DESCRIPTION

The block diagram of the transceiver is shown in Fig 1. The VFO is common to transmit and receive.

On receive, the incoming signal is passed from the aerial socket by the changeover relay to the band-pass filter (which greatly attenuates out-of-band signals) and then to the mixer where it is mixed with the VFO signal to produce an audio beat note. Most operators favour a note of 800 – 1000Hz, so the VFO will be offset by this amount. Because there will be many in-band but unwanted signals at the mixer input, there will be many more at the output after the mixing process. Although most of these will be beyond the range of hearing, some may be strong enough to overload the audio stages.

AN 80m TRANSCEIVER

The low-pass filter will all but remove those at radio frequencies, and also attenuate the higher audio frequencies. The audio preamp now amplifies the wanted signal to a sufficient level to drive the output stage. On transmit, the VFO signal is amplified by the driver and then the PA to give an output in excess of 1W. Harmonics are attenuated by the low-pass filter and the signal is passed to the aerial socket by a change-over relay. Keying is by means of a 'keying switch' (electronic, not mechanical) which keys the 13V supply to the PA. This keyed 13V also supplies the sidetone oscillator, the output of which passes to the audio output stage via the sidetone level control.

The RIT in this design allows the operator to tune approximately ±1.5kHz of the frequency on receive. The position of the RIT control does not, of course, affect the transmit frequency.

THE VFO

There is nothing remarkable about the VFO (Fig 2). Similar circuits are to be found in many pieces of equipment. The oscillator transistor is a 2N3819 field-effect transistor (FET). This operates at a low power level so the generation of excess heat, the enemy of VFO stability, is minimised. The supply is stabilised at +5V by IC1. The buffer stage consists of two BC108s, chosen because they are cheap, common, and I have a bag full of them. Most amateurs must have a few lying around. The coil L1 is wound on a T50-2 toroidal core. Wound with the number of turns specified, you should have a VFO that works on the correct frequency, which might not be the case if any old former and core were used (I don't know of any current source of new coil formers). In practice, frequency stability is quite adequate.

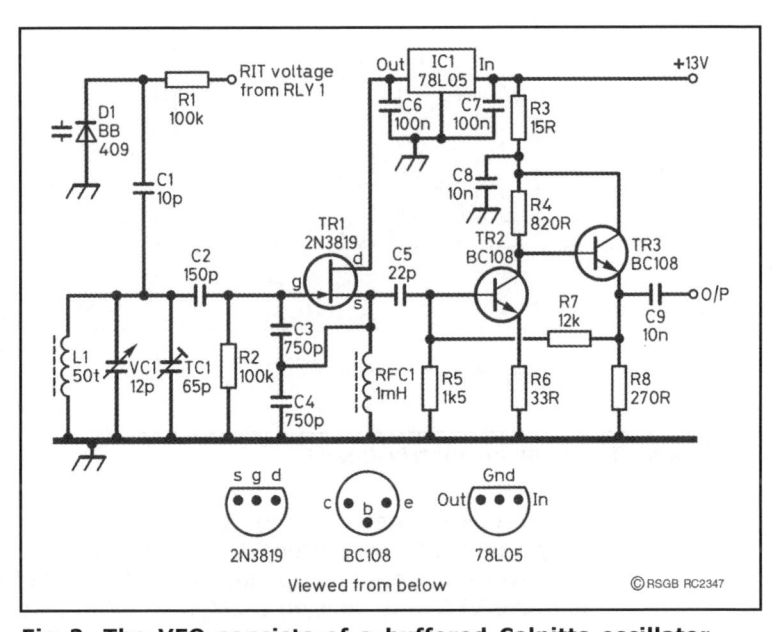

Fig 2. The VFO consists of a buffered Colpitts oscillator.

Mention must be made of the tuning capacitor, VC1. The tuning range is determined by the value of this component. 12pF allows coverage of the whole of the CW segment with a little to spare at each end, while 10pF will just cover the required range. As I prefer to have a little overlap, I used 12pF. With the amount of bandspread that this provides, the tuning rate is comfortable without a slow-motion drive. This, of course, simplifies mechanical construction. The value of a variable capacitor can be reduced by removing some of the plates.

WEEKEND PROJECTS

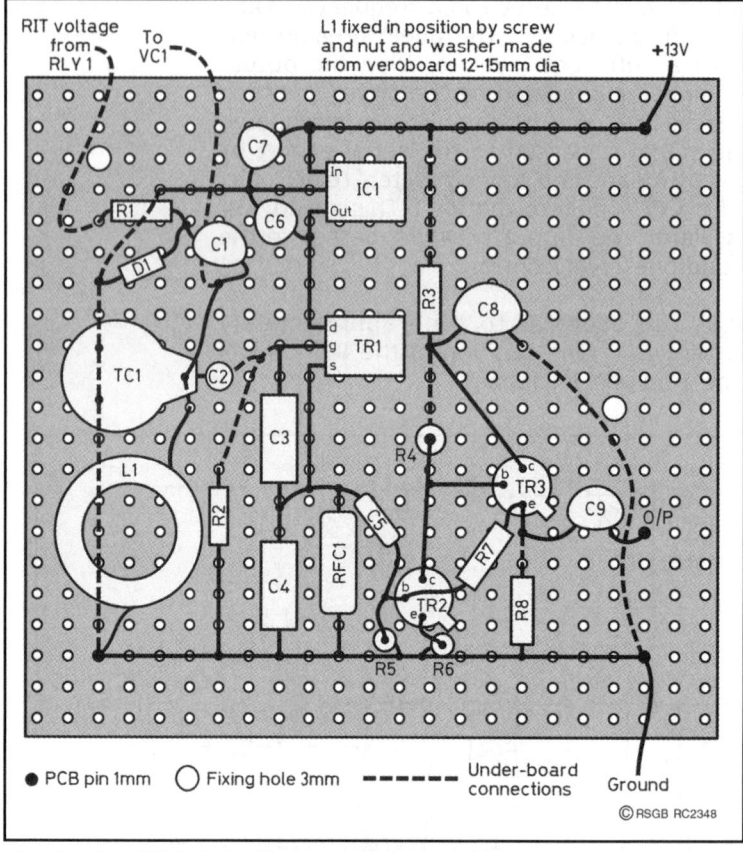

Fig 3. The VFO is built on matrix board.

Actually, it is the number of gaps that determines the value, so if you divide this number into the value, you will know how many picofarads each gap contributes. The excess plates should be carefully removed with long-nosed pliers, taking care not to damage the remaining ones or over-stress the bearings or shaft. So, if you have a component with a value that is a little too high, you can still use it, although I wouldn't recommend trying to reduce the value by more than 50% as the results become less predictable.

It was decided *not* to build the VFO 'dead bug' fashion for two reasons: (a) the rigidity (very important in a VFO) wouldn't be as good as it could be; and (b) thermal stability. With the frequency-determining components in contact with the chassis, they would be more prone to sudden changes in temperature than if they were mounted independently. As printed circuits had been ruled out, the VFO was built on a $2^1/_4$in × $2^1/_2$in piece of 0.1in matrix board. This, in turn, was mounted using two 12mm M2.5 screws, with extra nuts as spacers to keep the board clear of the chassis. See Fig 3 for the layout of the VFO board.

It turned out to be essential that the VFO be fully enclosed, not for the obvious reason of RF screening but because the slightest draught caused unacceptable drift. Holes are drilled in the screens to allow connections to pass through. Obviously this should be done before assembly. Initially the lid can be left off. Decoupling capacitor C40 should be mounted close to where the lead carrying the RIT voltage enters the enclosure.

THE RECEIVER
Referring to Fig 4 and the detail in Fig 5, the band-pass filter comprises T1, T2, C10, C11 and C12. The transformers are from Toko, which make circuits such as this easily reproducible.

AN 80m TRANSCEIVER

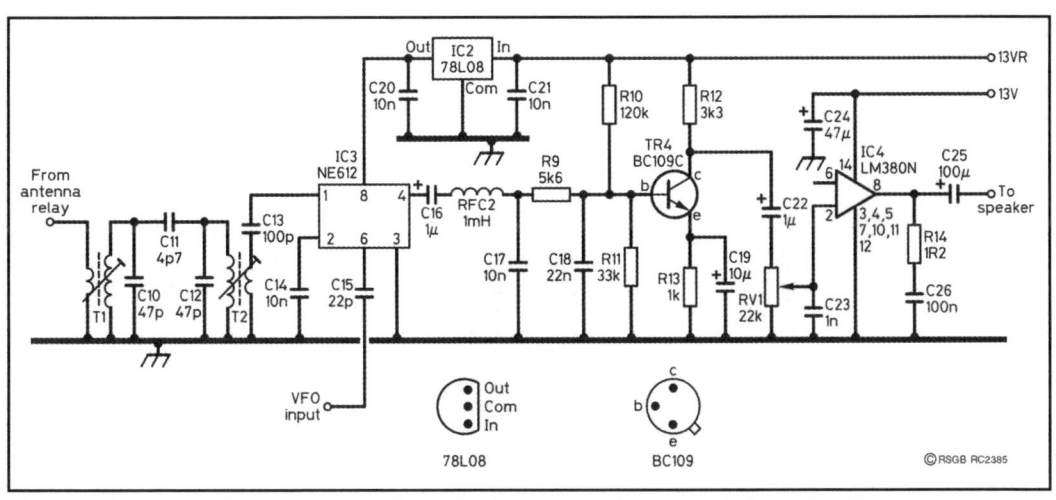

Fig 4. Circuit of the receiver. The VFO input is via a short length of miniature coax from C9.

Next comes the mixer. The NE612 is a useful device, containing an oscillator as well as a mixer. I decided not to make use of the oscillator in this case. The NE612 operates from an 8V supply provided by IC2, a 78L08.

As mentioned previously, the output contains many signals, audio and RF. The RF signals are unwanted and are removed by RFC2 and C17. The higher audio frequencies are progressively attenuated by R9 and C18, which form a low-pass filter. This gives 3dB attenuation at 1.3kHz and 6dB per octave thereafter. TR4 is the audio preamplifier. A BC109 was chosen rather than a BC108 because it gives more gain, which is required here.

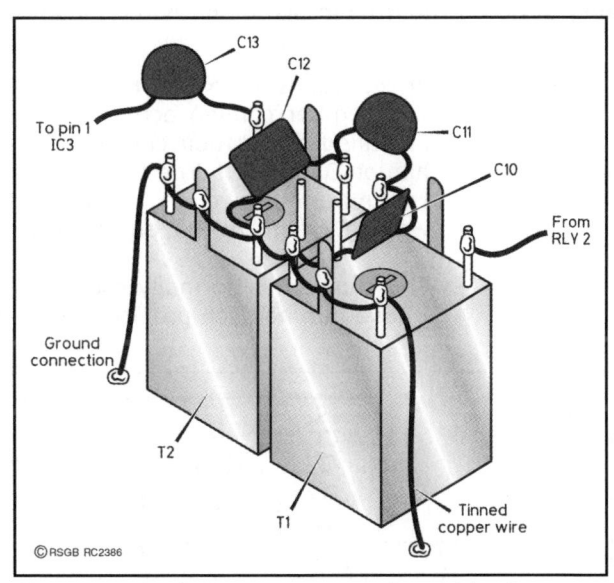

Fig 5. T1 and T2 are mounted upside down on the earth plane and wired as shown.

The audio output stage is an LM380N. As shown in Fig 6, it is mounted upside down in the 'dead bug' style.

Unlike the previous stages, it is supplied with 13V on transmit as well as on receive, as it is needed to amplify the sidetone signal to loudspeaker level.

TRANSMITTER
Referring to Fig 7, when the key contacts are closed, RLY2 switches the aerial to the PA stage and 13V from the receiver to the transmit

WEEKEND PROJECTS

circuits. The RIT control on the front panel becomes no longer operative, and the VFO offset voltage now comes from RV3.

The VFO signal is fed to the input of the driver stage, a BFY51. The collector is coupled to the PA by L2, a toroidal transformer. It is more usual to tune this with a trimmer but (on 80m at least) the tuning is very flat, so the number of turns is optimised for a standard-value capacitor, in this case 150pF (C29). There is adequate drive over the whole of the tuning range.

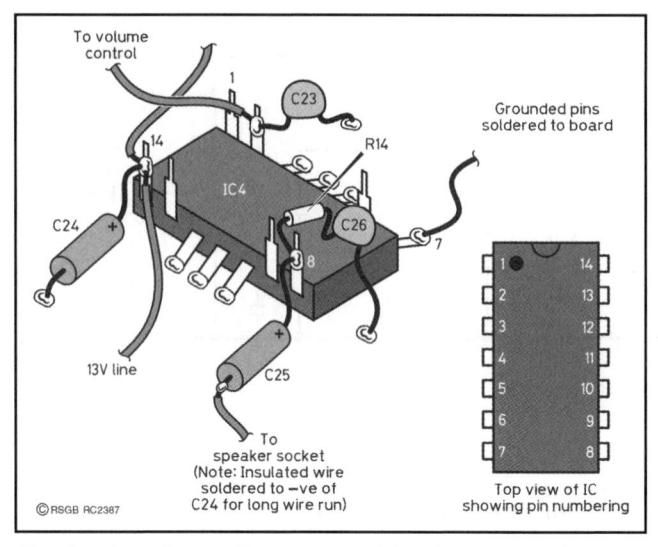

Fig 6. IC4, the audio output chip, is mounted 'dead bug' style on the ground plane and wired as shown.

Low-frequency parasitic oscillations, normally below 100kHz, sometimes occur in the driver stages of QRP transmitters, and can go unnoticed by the operator. I remember coming across a rough CW signal on about 3630kHz. I identified the station concerned, tuned down the band and also found him on 3560kHz. The two signals were 70kHz apart and, as expected, I found another rough CW signal at 3490kHz. The station concerned was only running 3W and was nearly 200 miles away.

This problem is normally caused by poor decoupling and is easily cured. I have had no such problems with this design. The output from TR5 is taken via a link winding on L2 to the base of TR6, the PA. This is a BFY51 and produces in excess of 1W output. The DC feed choke, RFC3, is not critical.

Some variation in wire thickness and number of turns shouldn't cause any problems, so just try what you have available. Take care, though, not to damage the enamel insulation when passing the wire through the bead, as this is easy to do.

Fig 7. The transmitter produces over 1W output and, thanks to adequate decoupling, is quite stable.

AN 80m TRANSCEIVER

The output is coupled by C32 to the low-pass filter, comprising L3, L4, C33, C34 and C35. It then passes via the changeover relay RLY2 to the aerial socket.

TR7 is the keying switch. Its use is preferable to directly-keying an RF stage such as the PA or driver, which can be unpredictable because RF has a tendency to find its way into keying lines. Also, there is the added advantage that the key or keyer only has to cope with the base current of TR7, a few milliamps.

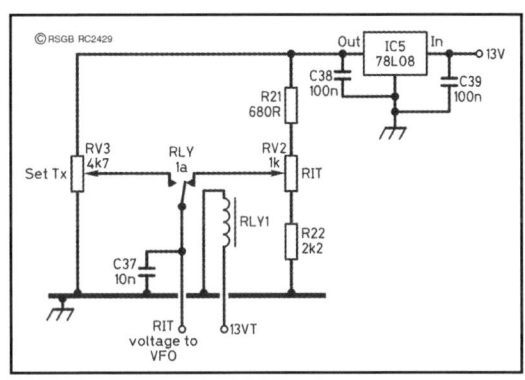

Fig 8. How the RIT voltage to the VFO is switched between transmit and receive.

The disadvantage of this kind of circuit is that it appears to be very good at injecting RF into supply lines and causing a shift in VFO frequency on 'key down' – something which is all too common in simple QRP equipment. The solution, again, is adequate decoupling, which is taken care of by C30 and C31.

RIT

The requirement here is that the VFO frequency can be varied over a small range without moving the main tuning control. As you can see from Fig 8, RV2, the RIT control, is active on receive only. On switching to transmit, the RIT control has no effect and the frequency reverts to that set by the main tuning.

The actual shift in frequency is accomplished by a BB409 varicap (variable capacitance) diode coupled to the VFO tuned circuit. When this is reverse-biased (ie positive to the cathode) it exhibits a capacitance which is dependent on the voltage so, by varying the voltage, it will tune the VFO over a small range. A fixed, stable voltage is required on transmit. This is provided by RV3. On receive, a variable voltage is provided by RV2, the front-panel-mounted RIT control. The supply to RV2 and RV3 is stabilised by IC5. Relay RLY1 selects the voltage from RV2 or RV3, depending on whether the transceiver is in the receive or transmit mode.

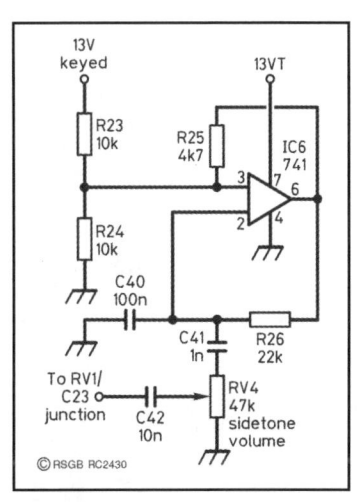

Fig 9. The sidetone oscillator and its associated level control.

SIDETONE

The sidetone oscillator (Fig 9) uses a 741 IC. It draws its supply from the 13V line, but is only activated on 'key down' by 13V from the keying switch, TR7. The output is taken to the volume control, RV1, via the sidetone level control, RV4, which is used to reduce the level of signal reaching the volume control. This is for operator comfort.

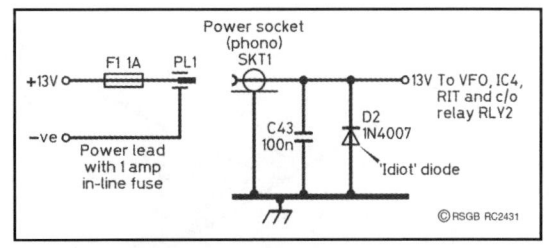

Fig 10. The connection to the supply contains protection against power of the wrong polarity being applied accidentally.

WEEKEND PROJECTS

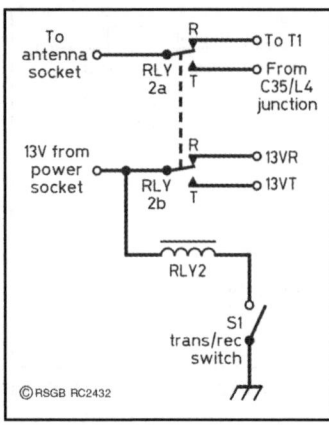

Fig 11. Transmit / receive switching.

POWER SUPPLY

Power is supplied to the transceiver via a phono socket on the rear apron (see Fig 10). This is decoupled by C43. There is a diode, normally reverse-biased, connected directly across the supply input. This is for protection so, if the supply is accidentally connected the wrong way round, the diode will be forward-biased and will conduct, blowing the 1A fuse, which should be installed in a holder in the power lead.

AERIAL SWITCHING

This is accomplished using a double-pole changeover relay RLY2 (see Fig 11).

MECHANICAL CONSTRUCTION

As stated previously, this project was designed for easy construction. Mechanical work has been minimised; nonetheless some is required. The copper-clad board should be double-sided glass fibre but SRBP could be used. This would be easier to cut, but the end result is not as rigid. If you have, or if you know someone who has, access to a workshop guillotine, this part of the project could be very easy.

If not, you will have to cut the board by hand. Possibly the best way to do this is clamp it in a vice between two pieces of angle iron which are then used as a cutting guide.

Using a hacksaw, cut the pieces very slightly oversize, then file down to the required dimensions. Refer to Fig 12 and Fig 13 for the dimensions.

When you have cut the base, front panel, rear apron, VFO enclosure and the two triangular supports, the holes can be drilled. First, mark their locations accurately and use a centre punch.

Start by drilling all holes 3mm in diameter. The larger holes can then be drilled to

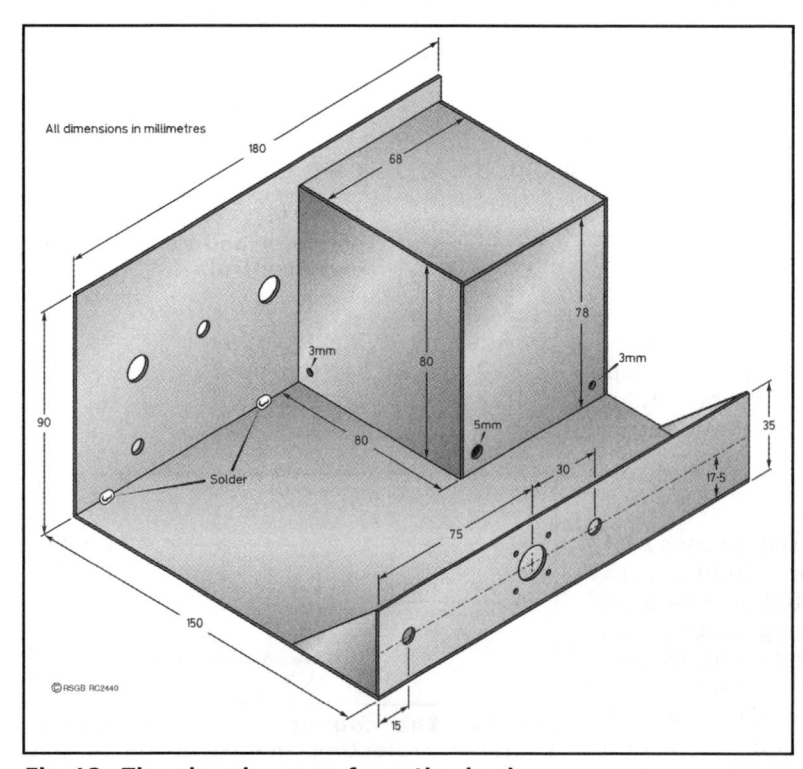

Fig 12. The chassis, seen from the back.

54

AN 80m TRANSCEIVER

6mm. At 10mm, more care is needed, as the drill can easily bite into the PCB. If you can, clamp the board firmly and drill through as slowly as possible.

The aerial socket can be any type you choose. If you want to fit an SO-239, a 16mm hole will be needed. This can be made by marking out, drilling to 10mm and enlarging with a file. Alternatively, drill a series of small holes (say, 2mm) inside the circumference of a 16mm circle, cutting out the centre and finishing off with a file.

Finally, drill the two 3mm holes in the base for the screws which support the VFO board. In this instance, there is no need to measure – just place the matrix board that you will be using in position, and mark through the holes.

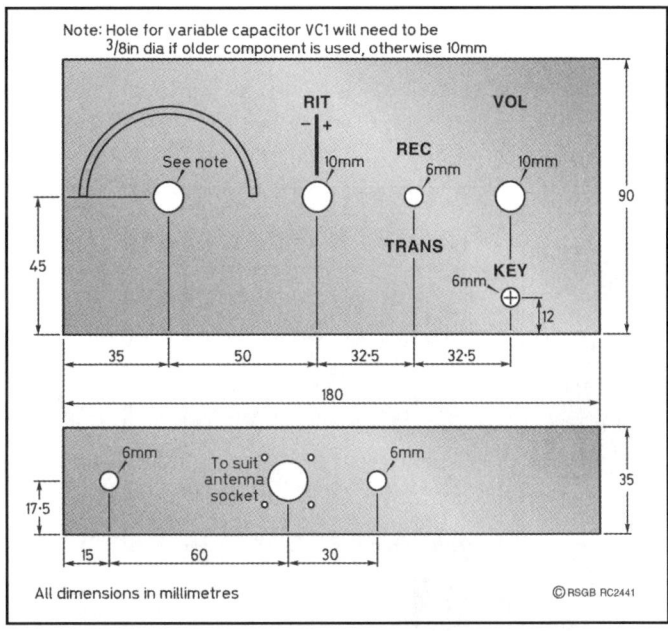

Fig 13. Front panel and rear apron layouts.

ASSEMBLY

With everything cut out and drilled, assembly can begin. Start by mounting the front panel on the base. Initially, use just one blob of solder in the centre. Then fix the side pieces of the VFO enclosure, again soldering lightly. After this, solder the rear apron in place. Finally, solder the two triangular supports in place. Do not mount the rear of the VFO enclosure at this stage. Inspect your work and if happy apply more solder to the joints and also solder in more places.

The end result will look more pleasing if the front panel is faced with card or thick paper, bearing labelling for the controls, along with a tuning scale for the VFO, and zero and ± marks for the RIT control. This can be done at any stage and fixed with adhesive.

ELECTRICAL CONSTRUCTION

This is best done in a logical order, rather than haphazardly. Refer to Fig 14, and start by mounting the large components: controls, sockets, switches, relays (the latter can be fixed in place with a spot of Super Glue®). Draw pencil marks on the base, along the line of the receiver (30mm from the side edge) and transmitter (30mm from the back edge). These are used as guides for the smaller components.

The first major job is the VFO. Fit all components and make all connections, taking great care to get the pin connections of the semiconductors correct. Check and double-check everything. Mount the VFO board, ensuring that the connections on the underside are

WEEKEND PROJECTS

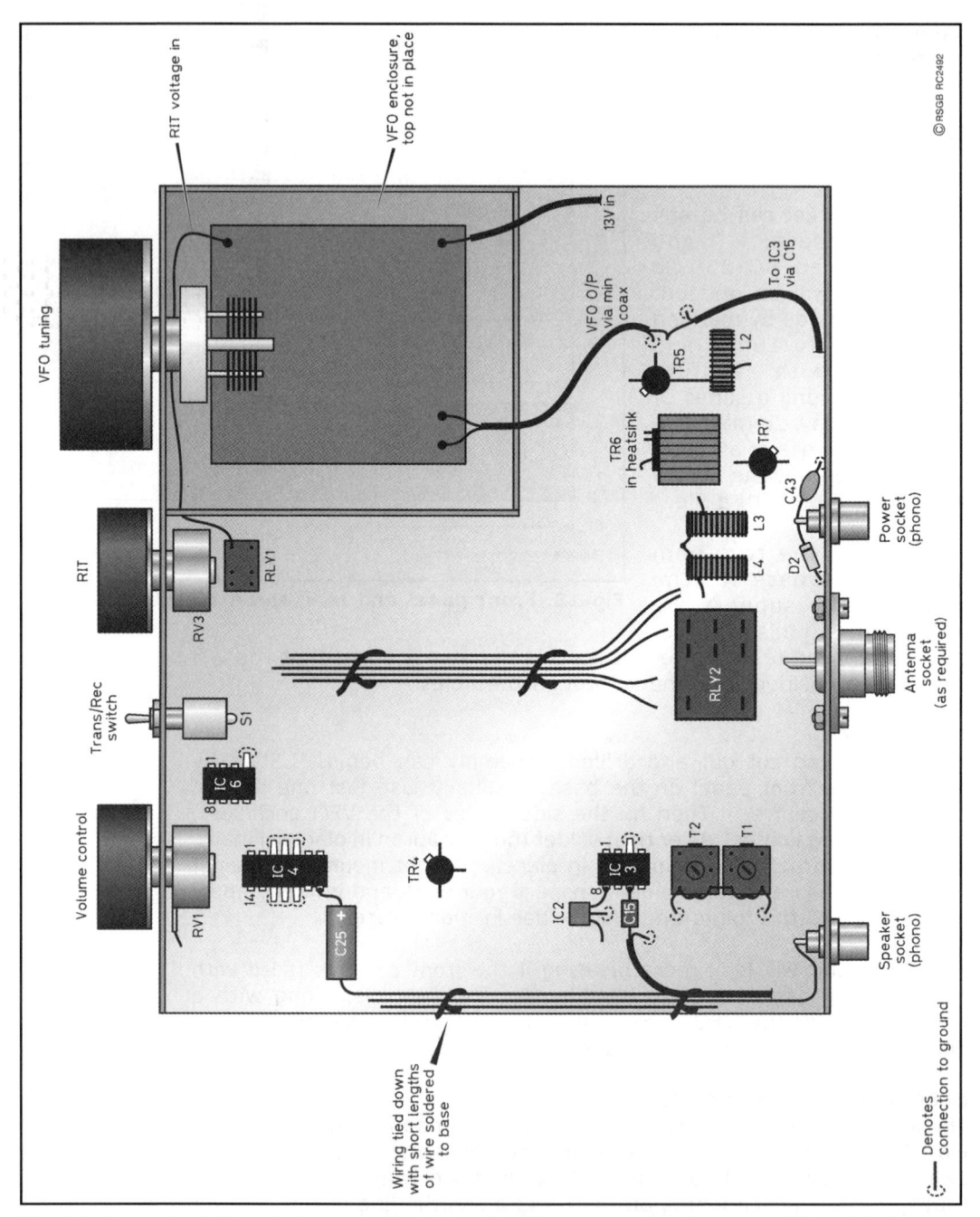

Fig 14. Layout of the major components. Smaller components are soldered between the larger ones.

clear of the base. Connect the variable capacitor using stiff wire (1.25 or 0.9mm). The lead to the RIT pin can be grounded for now. Connect a 13V supply and check that there is 5V at the output of IC1. All being well, the tuning range can now be roughly set. With VC1 at half-

AN 80m TRANSCEIVER

mesh, connect a short length of wire to the output (C9). Using a test receiver tuned to 3550kHz, adjust TC1 until a signal is heard. This is all that is needed at the moment. Don't be concerned if the VFO is not very stable at this stage.

Now for the receiver. Referring to Fig 4, start by assembling the band-pass filter (Fig 5) and fix it in place. Continue with IC3, IC2 and the smaller components. IC3 is mounted in the same manner as IC4 ('dead bug', ie legs upwards), but only pin 3 is grounded. Where grounding is necessary, leads are soldered directly to the base. Carry on until the receiver is complete. The volume control is connected with miniature screened audio cable, and the output from IC4 is connected to the loudspeaker socket. The receiver can now be tested using a temporary connection from the VFO to IC3, with 13V to IC4, TR4 and IC2, and a length of wire as an aerial to the input of the band-pass filter. Sensitivity will be low, but it should be possible to hear something by tuning the VFO up and down its range.

Top view of the completed transceiver.

Continue with the auxiliary circuits – the RIT, sidetone, changeover relay RLY2 [1] and the power connector. Permanent power connections can be made to all stages, a length of miniature coax connected to C15, and the wiring tied down neatly as shown. Now carry on and complete the transmitter and make a thorough check of everything.

TESTING

If at any stage a problem appears, turn off the power and investigate it. Connect a loudspeaker and an aerial, and turn on. Advancing the volume control should result in noise from the loudspeaker. Set the RIT control to zero. Tune the VFO and search for a steady signal at the centre of the tuning range. When one is found, using a suitable tool, adjust the cores of T1 and T2 for maximum volume. Now check the function of the RIT control.

Disconnect the aerial and connect a Morse key and a dummy load [2]. Set RV3 to the centre of its travel. Switch to transmit and press the key. The sidetone should be heard in the speaker. Holding the key down, tune the test receiver until the signal is found, somewhere near 3550kHz. If you have a power meter, it can be connected between the transceiver and dummy load. It should show an output of at least 1W.

SETTING UP

There are only three adjustments to be made – the VFO, RIT and bandpass filter – and they have already been roughly set.

WEEKEND PROJECTS

VFO
Close the vanes of VC1 fully. The tuning knob should be set to exactly nine o'clock. Now set the test receiver to 3490kHz (if VC1 is a 12pF component) or 3500kHz if it is 10pF.

Adjust TC1 until the signal is heard. Set the tuning control to the three o'clock position and check the frequency of the VFO by finding the signal with the test receiver. If using 10pF for VC1, the frequency should be about 3600kHz; if using 12pF it should be about 3610kHz. The VFO enclosure can now be completed, but don't use too much solder on the lid – you might want to remove it at some future time. The VFO must be calibrated, but not yet. Since you have just heated everything up with a soldering iron, now is not a good time – allow several hours to elapse first.

RIT
Set the VFO to the centre of its range, the RIT control to zero (centre), and tune the VFO to give a beat note on the test receiver. Switch to transmit and adjust RV3 for exactly the same note. Switch between transmit and receive and carefully adjust RV3 so that the note doesn't change.

Bandpass filter
With the aerial connected, find a steady signal in the centre of the tuning range and carefully adjust T1 and T2 for maximum signal strength. This time take care and make sure it is right.

Calibration
It just remains to calibrate the VFO scale and the transceiver will be ready for use.

Before using it in earnest, though, it would be a good idea to get a local amateur to listen to your signal to make sure that all is well.

AND FINALLY...
The transceiver is delightfully easy to use. To net onto a station, start with the RIT at its centre position and adjust the main tuning for zero beat. The RIT can now be set to give the desired note.

With a reasonable aerial (eg a G5RV) you shouldn't have any shortage of QSOs. You are unlikely to achieve WAC or DXCC (please prove me wrong) – on the other hand, you will almost certainly have no problems with TVI or BCI. Have fun!

NOTES
[1] As a particular relay is not specified for RLY2, you will have to work out the connections for yourself. If no data are available, this can be done visually and confirmed with a test meter on a resistance range.
[2] This can be three 150Ω 0.5W resistors in parallel.

AN 80m TRANSCEIVER

Resistors
R21	680
R22	2.2k
R23, 24	10k
R25	4.7k
R26	22k
RV2	1k
RV3	4k7
RV4	47k

Capacitors
C37, 42	10n ceramic disc
C38–40, 43	100n ceramic disc
C41	1n ceramic disc

Semiconductors
IC5	78L08
IC6	741
D2	1N4007

Miscellaneous items
RLY1	12V single-pole change over (SPCO) relay
RLY2	12V double-pole change over (DPCO) relay
F1	1A fuse with in-line holder
S1	SPST switch
PL1	Phono plug
SKT1	Phono socket

Table 1. RIT, sidetone, switching and power input circuits parts list.

Resistors - All resistors 0.6W metal film
R15, 20	1k
R16	220
R17	47
R18	56
R19	33

Capacitors
C27, 28, 30 – 32	10n
C29	150p
C33, 35	820p
C34	1500p
C36	2µ2

Inductors
L2	42 turns of 0.375mm enameled wire on a T50-2 core, plus a 5-turn link of PVC-covered wire over the main winding.
L3, L4	22 turns of 0.56mm wire on T50-2 core
RFC3	20 turns of 0.19mm wire on ferrite bead

Semiconductors
TR5, 6	BFY51

Table 2. Transmitter parts list.

WEEKEND PROJECTS

Resistors - All resistors 0.6W metal film unless specified otherwise
R9	5k6
R10	120k
R11	33k
R12	3.3k
R13	1k
R14	1.2
RV1	22k log pot

Capacitors
C10, 12	47p ceramic plate
C11	4p7 ceramic plate
C13	100p ceramic plate
C14, 17	10n ceramic disc
C15	22p polystyrene
C16, 22	1µ, 16V electrolytic
C18	22n ceramic disc
C19	10µ, 16V electrolytic
C20, 21	10n ceramic disc
C23	1n ceramic disc
C24	47µ, 16V electrolytic
C25	100µ, 16V electrolytic
C26	100n ceramic disc

Inductors
RFC2	1mH
T1, 2	Toko KANK3333R

Semiconductors
TR4	BC109
IC2	78L08
IC3	NE612
IC4	LM380N

Table 3. Receiver parts list.

Resistors - All resistors 0.6W metal film
R1, 2	100k
R3	15
R4	820
R5	1.5k
R6	33
R7	12k
R8	270

Capacitors
C1	10p ceramic plate
C2	150p close tolerance polystyrene
C3, 4	750p close tolerance polystyrene
C5	22p polystyrene
C6, 7	100n ceramic disc
C8, 9	10n ceramic disc
VC1	10p or 12p (see text)

AN 80m TRANSCEIVER

TC1	65p (Maplin WL72P)
Inductors	
L1	50 turns of 32SWG on T50-2 toroidal core
RFC1	1mH
Semiconductors	
TR1	2N3819
TR2, 3	BC108
IC1	78L05
D1	BB409

Table 4. VFO parts list.

For the **best** selection of **Amateur Radio books**

Only £14.99
plus p&p

Circuit Overload
The bumper book of circuits for the radio amateurs
By John Fielding, ZS5JF

This is the book that all keen home constructors have been waiting for! *Circuit Overload* includes 128 circuit diagrams, complemented by an additional 89 other diagrams, graphs and photographs and is a unique source of ideas for almost any circuit the radio amateur might want.

Circuits are provided that cover wide range of topics. Chapters have been devoted to audio, metering & display, power supplies, test equipment and antennas. Chapters have also been included that cover the design of low-pass, high-pass and band-pass filters. Nor are valves forgotten, with two chapters of *Circuit Overload* devoted to this fascinating topic.

Circuits include a two-tone oscillator for correctly setting up SSB transmitters, a temperature controlled crystal oven, a speech clipper, a solar panel battery charger, a simple intercom system, an RF 'sniffer' device, a product detector for an SSB / CW receiver, a Morse code oscillator, a phase modulator for an FM transmitter, a novel idea for an 'el cheapo' signal generator, a noise figure meter, a nicad battery charger and cell rejuvenator, a mains frequency monitor, an inrush current limiter, a high-power PIN diode antenna switch, crystal oscillators, an audio notch filter, audio amplifiers, 455kHz, 10.7MHz and 21.4MHz markers, a variable voltage bench power supply unit, a very low power transmitter for 1.3GHz to assist in aligning a 23cm converter, and much, much more.

Throughout the book, circuits are presented in an easy to understand fashion and many can be inter-connected to make a more complex item if so desired by the reader. If you are interested in home construction this book provides simple circuits and advice for the beginner with more complex circuits for the more experienced.

Size 240x174mm, 208p   ISBN 190508620-2

Radio Society of Great Britain
Lambda House, Cranborne Road, Potters Bar, Herts, EN6 3JE Tel: 0870 904 7373 Fax: 0870 904 737

www.rsgbshop.org

RSGB SHOP

STATION ACCESSORIES

A switched attenuator 64
An audio filter 67
A low-voltage alarm for battery supplies 70
Dry battery tester 72
A direct-reading capacitance meter 76
Earth-continuity tester 80
A diode/transistor tester 84
Extending the use of your dip oscillator 87
Dual-voltage supply from one battery 90
A simple electronic keyer 94
Field-strength measurement 97
A frame aerial for HF 99
Computer-to-radio interfaces 101
A cheap and cheerful approach to keying 105
A useful audio level indicator 108
A 'loop' alarm 110
An L-Match ATU 112
A charger for nicad batteries 114
An op-amp tester 118
Optical communication 123
Packet radio principles 128
A portable power supply 132
An amplified RF probe 135
A signal injector 137
An audio-driven s-meter for dc receivers 140
A time-out unit for digital modes 145
A T-match ATU 148
A 1750Hz toneburst for repeater access 151
A colourful voltage monitor 155
Voltage regulation 159
A bi-directional wattmeter 162
A standing-wave indicator for HF 165

A SWITCHED ATTENUATOR

How many times has a signal been too strong for the experiment you wish to carry out? It could be from an oscillator on the bench or from signals from an aerial overpowering a mixer. This attenuator will solve those problems.

WHERE CAN I USE IT?
Applications other than those given above are if you are interested in DF and need to attenuate the signal when getting close to the transmitter, or if you have a problem with a TV signal being too strong and causing ghosting on other signals (cross-modulation).

ATTENUATORS
These problems can be averted by using a switched attenuator (or 'pad'). The times that we need attenuators occur far more often than first realised. When designing the attenuators, account must be taken of the distinct possibility of poor screening. There is hardly any point in designing a 20dB attenuator when the leakage around the circuit is approaching this value. It is also important to decide on the accuracy required. If it is intended to do very accurate measurements, the construction has to be impeccable, but for comparisons between signals it would be possible to accept attenuation values to a lower degree of accuracy. [A 'pad' is the name given to a group of components with a known attenuation. – *Ed.*]

The most useful attenuator is a switched unit where a range from zero to over 60dB in 1dB steps can be covered. This is not as difficult as it first seems because, by summing different attenuators, we can obtain the value we need. It takes only seven switches to cover 65dB. The seven values of attenuation are 1dB, 2dB, 4dB, 8dB, 10dB, and two at 20dB; these can be switched in or out at will. As an example, if 47dB were needed, switch on the two 20dB pads plus the 4, 2 and 1dB pads.

CONSTRUCTION
The prototype attenuator is shown in the photo and in Fig 1. It was constructed 20 years ago and it is still in regular use. It is housed in a box made from epoxy PCB material. The top and sides are cut to size and soldered into a box shape.

It is easier to cut the switch holes prior to making the box. After the box has been constructed, screens made from thin brass shim should be cut and soldered between the switch holes. Next, the switches are

A SWITCHED ATTENUATOR

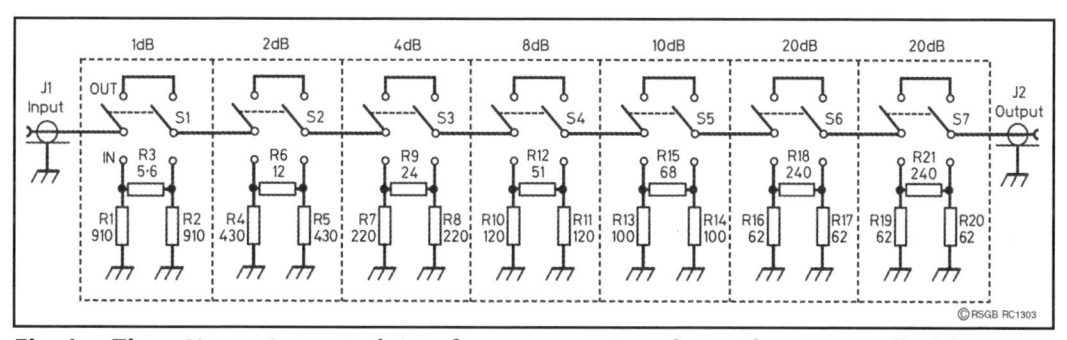

Fig 1. The attenuator consists of seven π-network sections, so-called because each pad (eg R1, R3 and R2) resemble the Greek letter pi (π). Input and output impedances are 50Ω.

fitted and the unit wired up. When this is done, the unit is checked and a back cover, securely earthed to the box, is fitted.

COMPONENTS

The switches must have low capacitance between the contacts and simple slide switches are the best selection. The Maplin DPDT miniature switch (FH36P) would be suitable and Maplin also supplies 1% resistors. Connectors to the unit must be coaxial but can be left to personal preference. The resistor values shown in Fig 1 determine the attenuator's accuracy at around 5%. This is done for practical reasons.

The completed attenuator.

WEEKEND PROJECTS

For example, if we wanted to make the attenuation value of the 4dB cell *exactly* 4dB, the resistor values would have to be 220.97Ω and 23.85Ω. You will see from Fig 1 that the values used are 220Ω and 24Ω, giving an attenuation value of 4.02dB.

PARTS LIST
Resistors All resistors in ohms, $\frac{1}{2}$W or 1W carbon or metal film

R1, 2	910
R3	5.6
R4, 5	430
R6	12
R7, 8	220
R9	24
R10, 11	120
R12	51
R13, 14	100
R15	68
R16, 17, 19, 20	62
R18, 21	240

Other items

S1 – 7	double pole miniature switch
J1, 2	SO239 sockets (or similar to suit your equipment)

AN AUDIO FILTER

This self-contained and inexpensive audio filter fits in the lead between a receiver and the headphones. Active audio filter circuits have been published in amateur radio journals over many years. Some are very elaborate and seem to have as many components as the receiver itself. Some are very clever, having several controls for adjusting the bandwidth and the frequency response, notch or peak tuning, and other functions. These tend to invite the operator to spend more time testing and driving the filter than operating the receiver! Read on and find out why this one is different.

ABOUT THE FILTER

This audio filter unit is elementary in concept and has only one control – a switch to turn it on and off and at the same time switch the filter in and out of circuit. Thus, instant checks can be made on the effectiveness of the unit. In these days of DSP, the inexpensive, simple, single-integrated-circuit audio filter still has a useful place in amateur radio. It can perhaps best be described as an *audio pass-band modifier*.

The filter is a very effective and useful accessory for CW and SSB reception. To keep it simple and to avoid adjustable controls, a 'most-often-used' response characteristic has been adopted, with a peak around 800Hz. This frequency was chosen principally for CW reception but is also usefully placed and shaped for SSB reception. It attenuates both the low and the high audio frequencies, narrowing the pass-band.

This narrower pass-band reduces the unwanted noise heard in the headphones. However, this filter cannot eliminate or reduce the image sideband from the direct conversion process used in simple receivers.

THE CIRCUIT

The active element of the circuit, shown in Fig 1, is an operational amplifier integrated circuit, using the common, inexpensive, general-purpose LM741, with a bridged RC filter connected in its feedback path. How it works is a very interesting academic study which need not concern us here. The RC filter comprises C2, C3 and C4, along with R4, R5, R6 and R7. The values of these filter components are important. Use good-quality components, preferably of identical type. The normal tolerance spread of component values results in a flat 'peak' in the response characteristic, making it acceptable for SSB and CW reception. Several filters have been built using store-bought components with the values shown. No special selection seems necessary. A 9V battery powers the unit. Remember to observe polarity and to switch it off when not in use.

WEEKEND PROJECTS

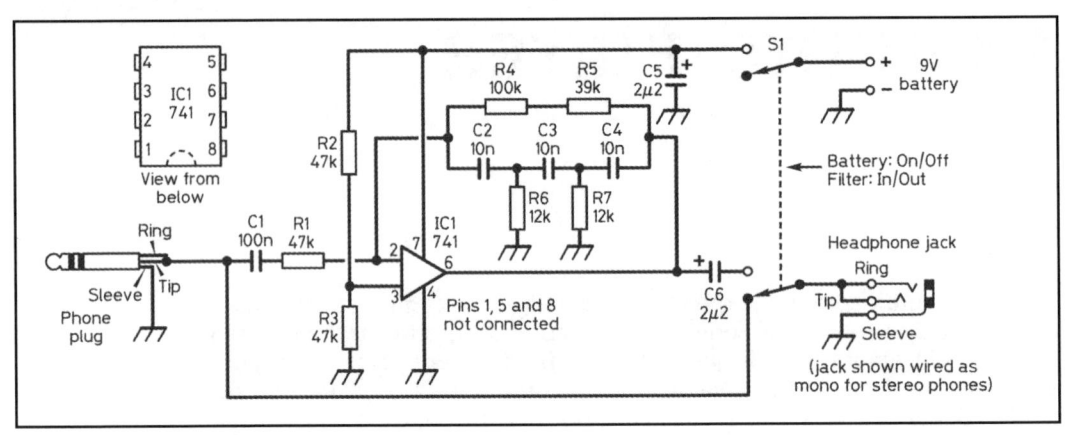

Fig 1. The simple audio filter. Note the pin assignment of IC1 shows the IC as it would be seen when wired-up 'dead bug' style.

CONSTRUCTION

The whole device fits into a plastic box. Select your own type! The double-pole changeover toggle switch is mounted on the lid of the box. The IC is glued upside down to a piece of printed circuit board which is bolted to the floor of the box. Wiring is by the 'dead bug' method. Note that IC pin numbers 1, 5 and 8 are not connected. Two pieces of tag strip hold the RC filter components and support other wiring junctions. The battery is held in place in the box by double-sided sticky tape.

There is nothing critical about the construction, and instability was not experienced. The whole unit can be built in an evening. It has a high 'go-first-time' factor. There is nothing that needs any setting-up but feel free to experiment!

TESTING

To plot a filter frequency response you need access to quality laboratory instruments with known specifications. Trial tests using lesser-grade 'test meters' have been found to be prone to measurement errors, due in particular to insufficient information being known about the frequency characteristics of the measuring apparatus used. Consequently, no simple procedure for testing with cheaper test gear is offered. However, it is easy and convincing to check the effectiveness of the unit by ear. Listen to a strong carrier signal, preferably with no modulation. Tune to what you estimate to be a typical signal tone or note for CW reception – about 800Hz.

Switch the filter in and out of circuit. Adjust the pitch of the received signal very carefully. You should find a position where the amplitude (loudness) of the wanted signal rises when the filter is in circuit. Ignore any change in the noise level, just listen to and refer to the change in the level of the received signal alone, as the filter is switched in and out of circuit. Above and below this frequency you should find that the loudness of the signal is higher when the filter is out of circuit, confirming a peak in the overall filter response. There is a very noticeable reduction in noise when the filter is in circuit.

PARTS LIST
Resistors –all resistors $^1/_2$W metal film
R1 – 3   47k, 5%
R4   100k, 1%
R5   39k, 1%
R6, 7   12k, 1%

Capacitors
C1   100n
C2 – 4   10n, 1%
C5, 6   2.2µ, 16V electrolytic

Semiconductor
IC1   LM741

Additional items
S1   DPDT toggle
Battery clip
PP3 battery
Case
Audio connectors (as appropriate)

A LOW-VOLTAGE ALARM FOR BATTERY SUPPLIES

Simple, direct-conversion receivers are prone to mains hum, and battery power prevents it. Assuming you use a rechargeable battery to power such a receiver, it should be recharged when the voltage falls below a predetermined level. Voltage checks can be made with a meter, but this can often be overlooked and the battery then becomes too flat to operate the equipment. This low-voltage alarm sounds when the voltage drops below a certain level to remind you that the battery needs to be recharged.

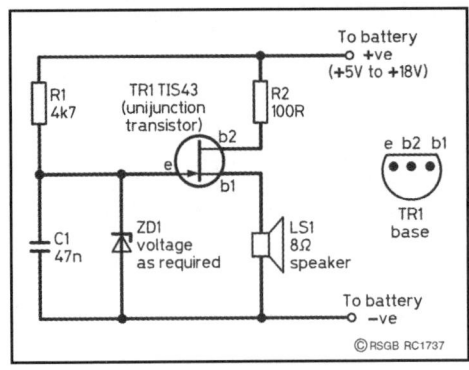

Fig 1. The low-voltage alarm uses a unijunction transistor as an oscillator.

OPERATION

The alarm, shown in Fig 1, is connected across the receiver when in use. It is basically an oscillator circuit designed around a unijunction transistor, TR1. The emitter of TR1 is biased by the resistor R1 and the Zener diode ZD1. The circuit will oscillate when the emitter has a sufficiently low voltage on it, and will cease oscillation if this voltage rises.

ZD1 must be chosen to determine the alarm voltage you require, and some figures for your guidance are given in Table 1. To obtain alarm voltages between the values given in the table, adopt the following procedure. Let us suppose you want an alarm voltage between 10.9 and 12.1V. Choose the Zener diode voltage that gives a 10.9V alarm, and insert a forward-biased silicon diode in series with the Zener, the cathode of the silicon diode being connected to the cathode of the Zener. This effectively increases the Zener voltage to 7.2V (ie 6.6V plus the 0.6V across the silicon diode). This will raise the effective alarm voltage to between 10.9 and 12.1V.

Zener voltage	Alarm voltage
6.6	10.9
8.2	12.1
12	19

Table 1. A guide to the alarm voltages produced by various Zener diodes.

The alarm draws a small amount of current all the time so, although it could be left connected permanently across the battery itself, this is not recommended as it would eventually run it down.

A LOW-VOLTAGE ALARM FOR BATTERY SUPPLIES

CONSTRUCTION

The circuit is built on a small piece of plain, single-sided PCB. Four saw-cuts are made through the copper, as shown in Fig 2, to make 'pads' to which the components are then soldered. When cutting, be careful not to cut right through the board, just the copper. I used a junior hacksaw to do this. The completed alarm can either be built into a case of its own – as I did – or into an item of equipment.

The PCB can be fixed to the back of the loudspeaker with adhesive pads.

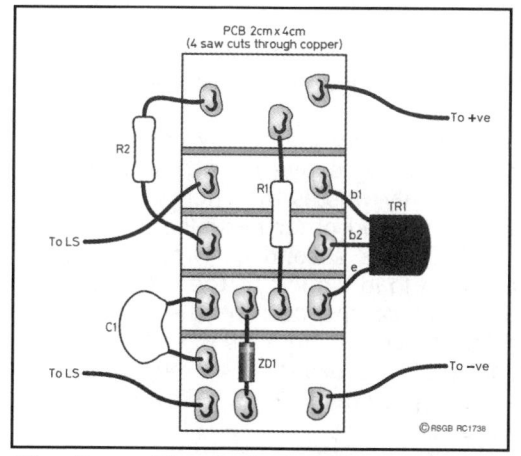

Fig 2. Physical layout. Before commencing assembly, make four saw-cuts through the copper of the PCB.

LIMITATION

The low voltage alarm relies on there being sufficient voltage in the battery you are monitoring to operate the alarm itself. Consequently, if the battery falls to such a low voltage that the oscillator will not operate, the alarm will not sound. Your receiver will not operate either, so you have two clues, both pointing to a flat battery!

PARTS LIST

Resistors
R1   4.7k, $\frac{1}{4}$W, 5%
R2   100, $\frac{1}{4}$W, 5%

Capacitors
C1   47n polyester

Semiconductors
ZD1   see text
TR1   TIS43

Additional items
LS1   8-ohm, 1W
Case to suit (if required)
PCB   2cm × 5cm

DRY BATTERY TESTER

The output voltage of a dry battery with no load, as measured by a digital voltmeter for example, will give an indication of the state of the battery. However, that is only part of the story – even a spent battery can show quite a high voltage when out of circuit but, as soon as an attempt is made to draw current, the voltage will fall dramatically.

You can check the current capability of a dry battery by connecting it directly across an ammeter but, as this is effectively short-circuiting it, it is not to be recommended. It not only wastes the energy of a good battery but can also cause overheating and other permanent damage if the short-circuit lasts for more than a second or two. A better method is to connect the battery to a load similar to the one it would experience in normal use and measure the current the battery is capable of providing. The device described here uses this method, and will test the current capability of the two most commonly-used dry cells, the 1.5V and 9V types, and provide a visual indication of the state of the battery via a series of LEDs.

HOW IT WORKS
For the test, the battery is connected to a load resistor capable of withstanding, without 'burning out', the current the battery can provide. The voltage dropped across the resistor, which is proportional to the current (by Ohm's law), is compared with a series of reference voltages chosen to represent the changeover points between the current ranges of interest. Where the generated voltage exceeds the particular reference voltage, one or more of a series of LEDs light to give an indication of how much current the battery is supplying and hence whether it is in an 'excellent', 'good', 'adequate' or 'needs replacing' state.

THE LOAD
AA- and PP3-type Duracell batteries will provide up to 2A for intermittent use. A suitable test load that draws sufficient current to test the battery's state without wasting too much of its power will be one that allows around 1A to flow.

THE CIRCUIT
This is shown in Fig 1. R1 and R2 are the load resistors for 1.5V and 9V batteries respectively, and switch S2(a) selects whichever test is required. The same value of current passing through a 1Ω and a 4.7Ω resistor will generate a different potential difference (PD) across each resistor of course, and so for the indicator part of the circuit to be

DRY BATTERY TESTER

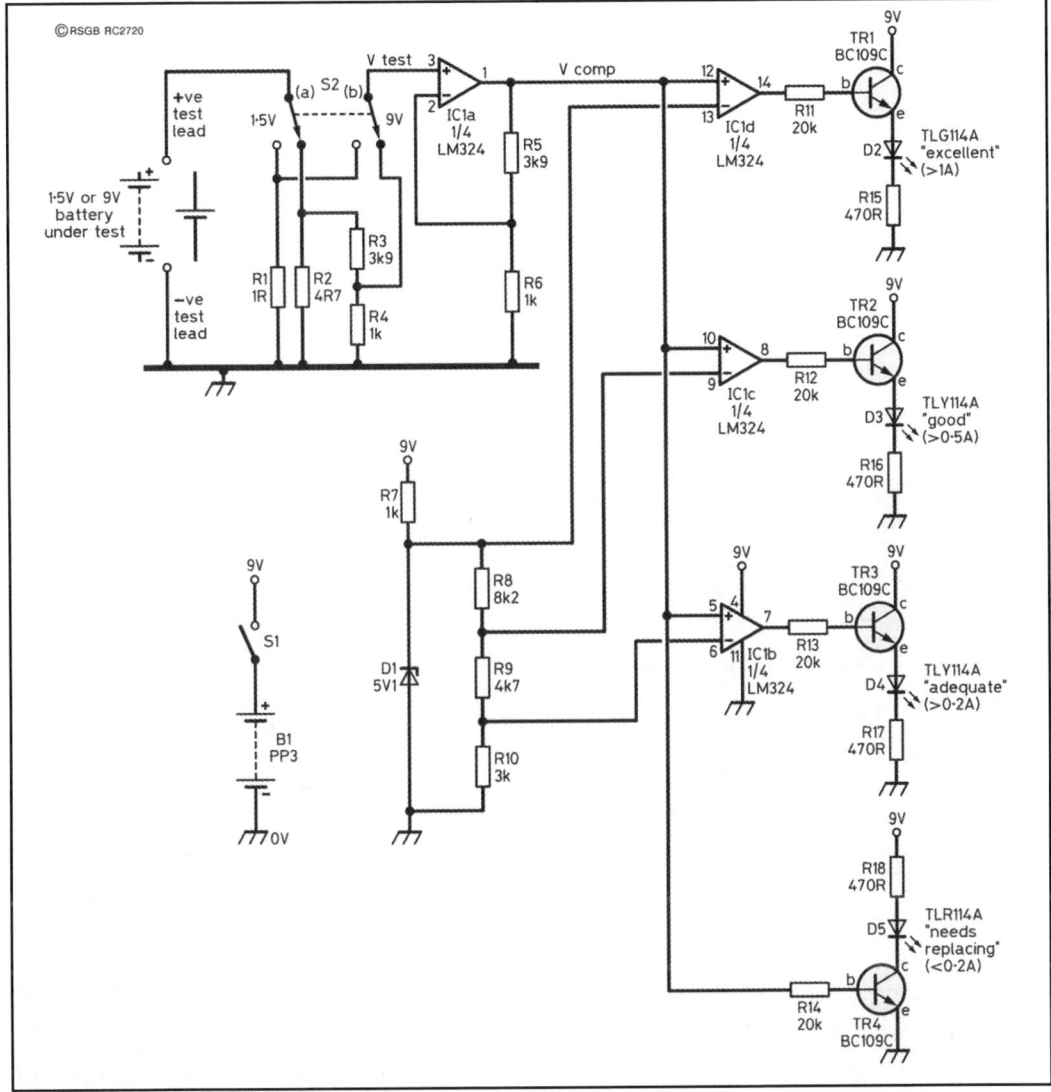

Fig 1. The unit works by checking the voltage that a battery is able to deliver when a substantial load is placed on it.

common to both tests, the PD across the 4.7Ω resistor is split by a potential divider consisting of R3 and R4. Using this method, the value of *V*test is nominally 1V/amp of current supplied by the battery under test for both types of battery. S2(b) selects between the non-divided and divided source voltage as appropriate.

Before passing to the comparator stage, *V*test is amplified by IC1a, connected as a non-inverting amplifier with a nominal gain of five. This means that the voltage applied to the comparators (*V*comp) will range from 0V to around 6V, and give more predictable switching of the comparators by avoiding using voltages around or below the turn-on voltages of the IC's internal p-n junctions (about 0.7V).

WEEKEND PROJECTS

The reference voltages against which *V*comp is compared are generated by a voltage divider network consisting of R8, R9 and R10 across a 5.1V Zener diode, D1. The use of a Zener diode ensures that the reference voltages remain stable, even if the power supply battery voltage changes with use. R7 limits the current through D1 to its working value. Reference voltages of 1V, 2.5V and 5.1V are used to give changeover points at nominally 0.2A, 0.5A and 1A.

Indication of current values of less than 0.2A (but greater than approximately 0.14A) are given by the LED driven by TR4. This part of the circuit takes further advantage of the turn-on voltage of a p-n junction, in that TR4 will not switch on until *V*comp is above around 0.7V. Hence there will be no false indication of a current of less than 0.2A when no battery is being tested, even if *V*comp is not exactly zero due to any small offset or bias currents associated with the op-amps. IC1b, IC1c and IC1d are op-amps used to act as comparators by not employing any feedback. Due to the high gain of op-amps used in this way, their output will be around 0V when the non-inverting (+) input is less than the reference voltage on the inverting (-) input. In this state, the transistor connected to its output (one of TR1 to TR3) will be 'off' and the associated LED (one of D2 to D4) will not light. As soon as *V*comp rises above the relevant reference voltage however, the op-amp output will immediately go to around 9V, switching 'on' the appropriate transistor and lighting the associated LED. R11 to R18 limit the current to the working values of the transistors and LEDs.

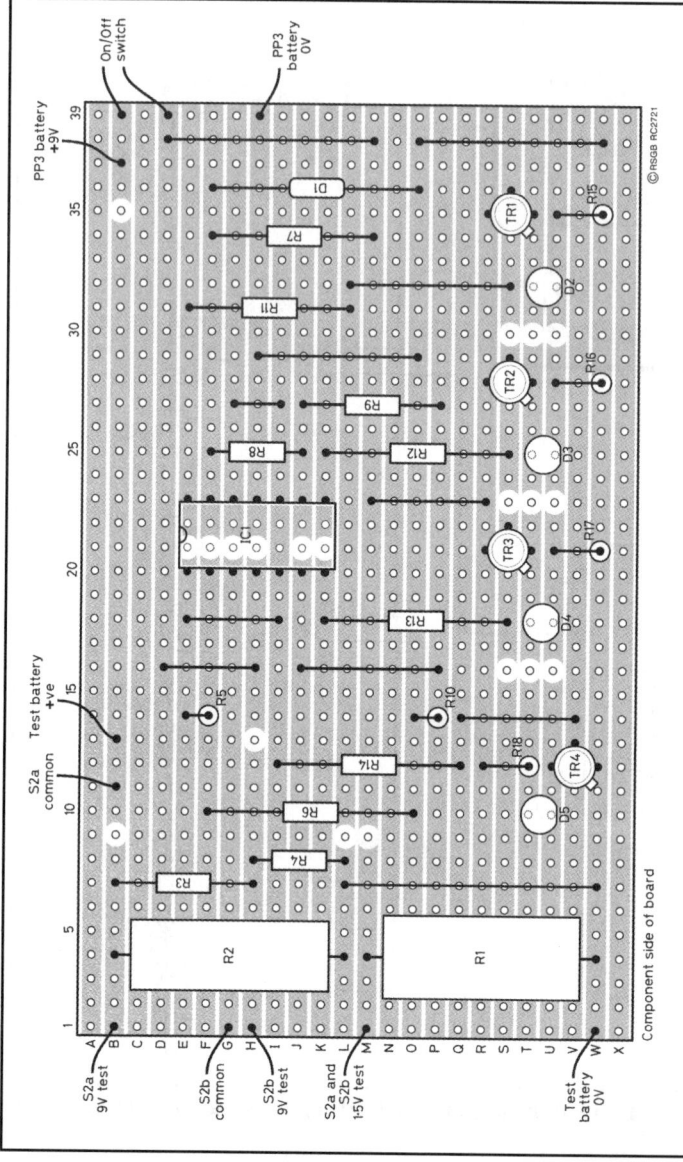

Fig 2. Stripboard layout of the dry battery tester.

DRY BATTERY TESTER

CONSTRUCTION

A suggested stripboard layout is shown in Fig 2. Construction is straightforward, the only particular point to mention being the correct orientation of the IC, transistors, Zener and LEDs. The markings that indicate their correct orientation are reproduced in Fig 3.

The components are all general-purpose low-power types, apart from the load resistors which ideally need to dissipate 5W (although 3W types should suffice as the current to be carried only passes for short periods). It is useful to employ different colours for the positive and negative test leads as the correct test-battery polarity must be observed for the LEDs to light correctly. Alternatively, solder battery cell holders and clips permanently to the ends of the test leads (if this is done only one test battery should be connected to any holder or clip at a time, and it should be removed immediately after testing).

BATTERY TESTING

With S2, select the 1.5V or 9V battery test as appropriate, switch on, connect the battery to the test leads and note which LEDs light.

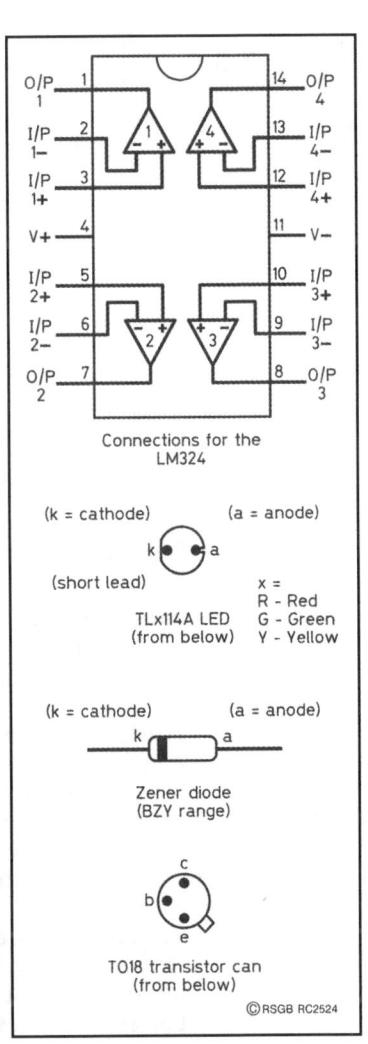

Fig 3. Orientations of the semiconductors.

PARTS LIST

Resistors - All resistors metal film, 0.6W, 1%, unless specified otherwise
- R1 — 1Ω, 5W, see text
- R2 — 4.7Ω, see text
- R3, 5 — 3.9k
- R4, 6, 7 — 1k
- R8 — 8.2k
- R9 — 4.7k
- R10 — 3k
- R11 – 14 — 20k
- R15 – 18 — 470Ω

Semiconductors
- IC1 — LM324
- TR1 – 4 — BC109C
- D1 — BZY5V1 0.5W Zener
- D2 — TLG114A green LED
- D3, 4 — TLY114A yellow LED
- D5 — TLR114A red LED

Additional items
- S1 — SPST
- S2 — DPDT
- Stripboard
- PP3 battery and clip
- Battery holders for test batteries, if required, see text
- Case to suit

A DIRECT-READING CAPACITANCE METER

Bags of surplus unused capacitors from rallies become an attractive proposition provided you can fathom out what they are. You may also possess polystyrene capacitors where the marking ink (used to show the values and working voltage) has rubbed off. This instrument saves you trying to recall myriads of capacitor value codings. You will find other uses, for example the ability to check the setting obtained with a variable capacitor used to find a resonance. The reading obtained by this direct-reading capacitance meter gives the value needed for a substitute fixed capacitor.

THE CHOICE
There have been various approaches made in designing capacitance checkers, including:

Multi-range test meter
A suitably-scaled multimeter can give approximate values for capacitors, using what is really one of the AC voltage test ranges in conjunction with a mains-powered step-down transformer. Such a solution is messy but reasonably effective. It is not a recommended practice with low-voltage capacitors or for use by beginners!

Test meter with internal AC source
Many modern digital multimeters have capacitance ranges. An alternating voltage is generated by an oscillator powered from internal batteries. These meters have their uses, though, but are expensive. However, the display format provided, which usually consists of three-and-a-half digits, is unsuitable for taking variable readings, as when checking a suspect plastic dielectric tuning capacitor.

Service technician's resistance / capacitance test bridge
Resistance/capacitance ratio bridges used to be one of the tools of every service technician. These versatile units often have additional features. There is usually a source of high DC voltage (0 to 500V variable) for the forming of high-voltage electrolytic capacitors. This voltage may also used for testing leaking paper capacitors which, as every service technician knows, contribute to the demise of resistors followed by valves and transformers.

Purpose-designed capacitance checker
A direct-reading capacitance meter has the facility to transfer capacitive reactance, in linear fashion, to the scale of an analogue direct-reading meter. The analogue meter is equally happy with fixed or variable capacitance readings.

A DIRECT-READING CAPACITANCE METER

The design of cheap and accurate direct-reading capacitance meters becomes a lot easier with the availability of solid-state integrated circuits. The simplest designs settled on the use of the 555 timer IC operated as a monostable oscillator.

The definitive work was done by A Wilcox for the magazine *Television* in the early 1970s. Constructional projects subsequently appeared in *Electronics Australia* October 1976, and in *QST* January 1983. The design used for this project begins with the *QST* circuit and then sets out to overcome many of the perceived limitations.

THEORY

An explanation is provided for those who like to know a little more about such things. To make the theory aspects clearer, please refer to the circuit given in Fig 1.

The 555 is used as a monostable oscillator and the test capacitor, which we will call Cx, is first charged and then discharged, and the meter indicates the average discharge current. The formula is

$$I_{ave} = \frac{V \times C_x}{(R_A + 2R_B) \times C_1} \times K ,$$

where V is the voltage to which C_x is charged and K is a constant, depending on the charge and discharge time of the 555 circuit, including the contribution made by the IC internal structure as well as the external resistance ratios and C1.

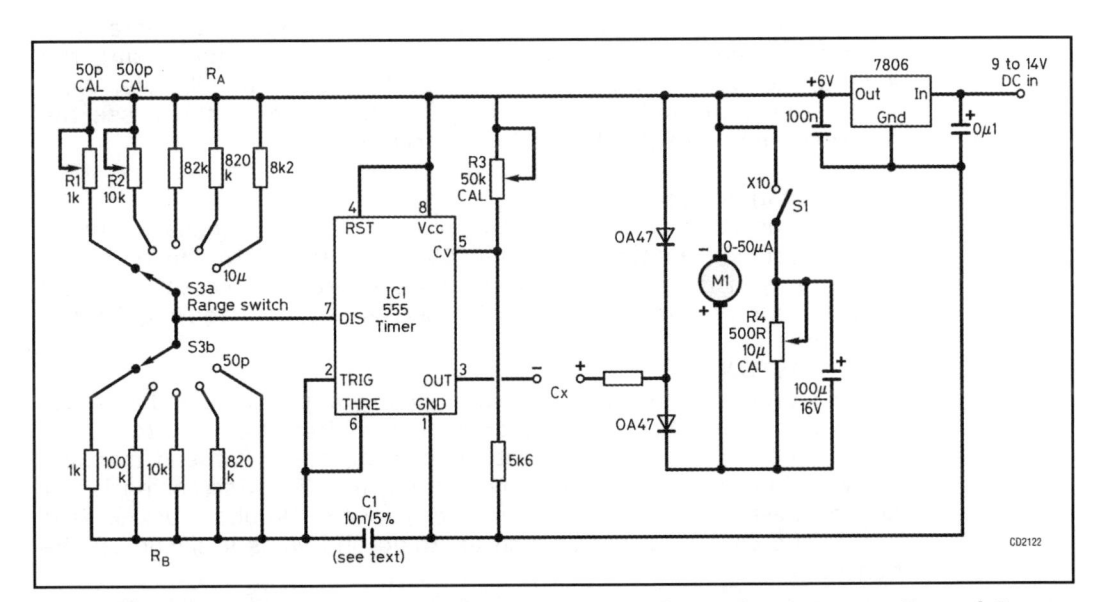

Fig 1. Circuit diagram of the capacitance meter. The resistor groups R_A and R_B are discussed in the text, as are the pair of resistors selected by switch S3. S3a selects R_A, and R_B is selected by S3b. All fixed resistors are 5% ¼W carbon types, and polarised capacitors are electrolytics.

CONSTRUCTION
Various approaches are possible but using the contents of the junk box involves little or no cost at all.

Front panel
The specification calls for a 50µA meter mounted on a plastic box. The bottom range reads 50pF full scale. A switch provides extra ranges. There are five rotary switch positions, extending the ranges by a decade factor each time.

Range extender
We now have, at position 5 of the switch, a full-scale direct reading of 0.5µF. At this point, the meter needle starts to flicker visibly at the 555 oscillator frequency. In order to extend the range further, a ×10 shunt is paralleled across the basic 50µA meter movement. This extends *all ranges* by a ×10 factor, and the maximum becomes 5µF. As the meter flicker is unacceptable, an auxiliary switch also brings in a 100µF smoothing capacitor across the meter movement.

The applied voltage is about 2V on the terminals, which is unlikely to harm even the most delicate components.

Construction details
Short internal wiring is preferable and the IC PCB is mounted right at the test terminals. Additionally, the 555 is mounted in a socket for peace of mind, allowing easy replacement should this ever be necessary.

Calibration
Use a good 0.047µF capacitor on range 3 to set the 50kΩ calibrator to produce a reading of 47µA. Then set the ×10 range (using the appropriate switch settings) with its preset to read 5µF full scale. Next adjust the 50pF and 500pF controls. You are now able to see the difference between 2.7pF and 3.3pF accurately.

Low-cost operation
The unit has a small mains power supply and needs a secondary winding able to provide anything between 9V and 14V DC to the regulator input. If you do not like to wire mains-operated accessories, a 9V alkaline battery can be used, but do not omit the voltage regulator.

CONCLUSION
There is great satisfaction to be had in building up reliable and accurate test equipment. Spend some time on the cosmetics; try to obtain matching knobs, give consideration to using brass flathead machine screws where they appear on the outside, and if silk-screening is out of the question you can use Dymotape® labels. Rubber stick-on feet ensure the case does not scratch anything and stops it sliding off the bench.

With a few known 2% calibration capacitors used from time to time as reference sources, this meter has been found to provide consistent

A DIRECT-READING CAPACITANCE METER

and quite accurate measurements. The prototype has been in regular use for several months and performs consistently well.

PARTS LIST
Numbered components not listed below are for text reference only.

Resistors
R1   1k PC-mounting
R2   10k PC-mounting
R3   50k PC-mounting

Semiconductor
IC1   555 timer
S1   SPST switch

Hardware
S3   2-pole 5-way rotary switch

EARTH-CONTINUITY TESTER

When using mains-powered electrical equipment, a good-quality protective earth system is very important for safety. Good earth connections are additionally important for radio operation, both for protection against lightning strikes and also for the greater effectiveness of aerials that use earth as one half of a dipole.

In situations where the earth path is a *functional* earth as opposed to a *protective* earth, a simple low-voltage, low-current continuity tester or resistance meter is usually sufficient for checking earthing resistance, but for a proper test of a protective earth a high-current tester is needed. This is because a deteriorating earth connection in the form of a stranded wire where many of the strands are broken will still show a low resistance to a low-current tester but, in a fault situation when the earth path needs to pass a high current to ground and thus trigger a protective device, the high current causes the remaining strands to 'burn out', ie go open-circuit, before the protective device has time to operate; the protection is then nonexistent.

SAFETY STANDARDS

Recognising this situation, the British and European safety standards for electrical safety, for example BS EN 60335-1 for household equipment, demand that the resistance of the protective earth path between an exposed metal part and the protective earth pin is less than 0.1Ω.

The equipment needed for checking to this standard is specialised and expensive, but this simple project provides a low-cost alternative and will check resistance at 2 – 3A if good-quality batteries are used. To simplify use, the circuit gives a pass / fail indication instead of a resistance value.

WHEATSTONE BRIDGE

The circuit can be considered in three parts; *test*, *detector* and *output indicator*. See the circuit diagram in Fig 1.

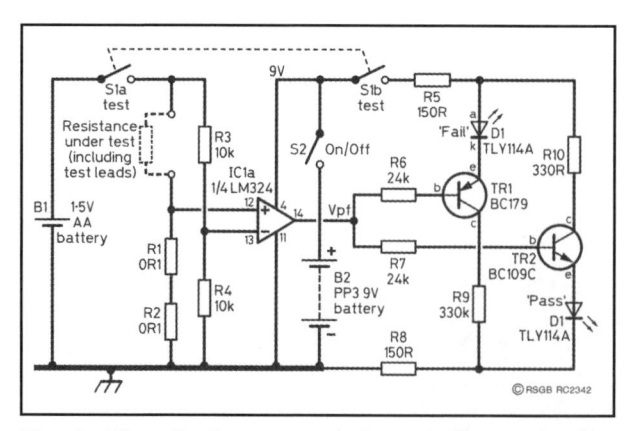

Fig 1. Circuit diagram of the earth continuity tester.

EARTH-CONTINUITY TESTER

The *test* part of the circuit is based on a Wheatstone bridge, where the earth resistance path forms one of the resistance 'arms'. See Fig 2 for the principle behind a Wheatstone bridge. As a consequence of the values of resistance chosen (the test leads are assumed to have a resistance of 0.1Ω), if the earth resistance is less than 0.1Ω, the voltage between the midpoints of the two halves of the Wheatstone bridge will be positive, and if it is less than 0.1Ω, it will be negative.

This is fed to the detector part of the circuit. The *detector* is an op-amp wired as a comparator. Connected in this way, it has such a high gain that its output is roughly equal to either the positive or negative supply rail voltage, depending on whether the PD between its non-inverting and inverting inputs is positive or negative. It doesn't matter whether the PD is large or small – the output will always be at either extreme. This means there will always be a definite pass or fail indication from the detector, no matter how large or small the output from the Wheatstone bridge. This is important, as it means correct operation of the circuit doesn't depend on the voltage of the high-current battery, particularly as it is a chemical type whose output voltage can fall dramatically when a high current is being drawn. The pass / fail voltage V_{pf} from the detector then passes to the output indicator circuit.

The *output indicator* circuit consists of two LEDs, driven by transistors to provide sufficient current, which indicate either a pass or a fail for an earth path resistance of less than or more than 0.1Ω. TR2 is an npn transistor which switches on when its input is high, while TR1 is a pnp type which switches on when its input is low. A separate supply voltage is needed for the op-amp and LED circuit, since the test battery voltage will drop under a heavy load current.

Because the output of the op-amp does not swing completely to the positive and negative supply rails, measures need to be taken to ensure that the LED driver transistors switch off correctly.

CONSTRUCTION
A suitable stripboard layout is shown in Fig 3, and Fig 4 shows how to identify and orientate several of the components. Use thick wire for the test leads!

HOW TO USE
Using flying leads with suitable connectors, eg crocodile clips, connect the circuit to each end of the earth path to be tested. This would usually be the

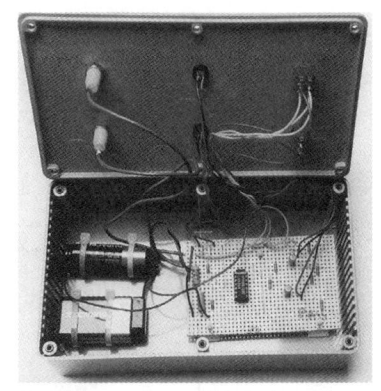

Inside a completed tester. Note that in a 'cased' project, the LEDs are removed from the stripboard and brought out to the front panel.

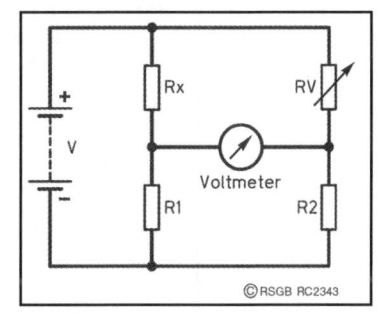

Fig 2. In a conventional Wheatstone bridge circuit, the value of an unknown resistance (Rx) is determined by adjusting a variable resistor (RV) with a calibrated scale until the reading on the voltmeter is zero. At that point (Rx/R1) = (RV/R2) so the value of Rx is then given by Rx = (R1 × RV)/R2. The advantage of this method is that, at the balance point, no current flows through the voltmeter, so the resistance of it doesn't affect the measurement. This is a sensitive method for detecting small changes in resistance, as a small change causes a large meter reading.

WEEKEND PROJECTS

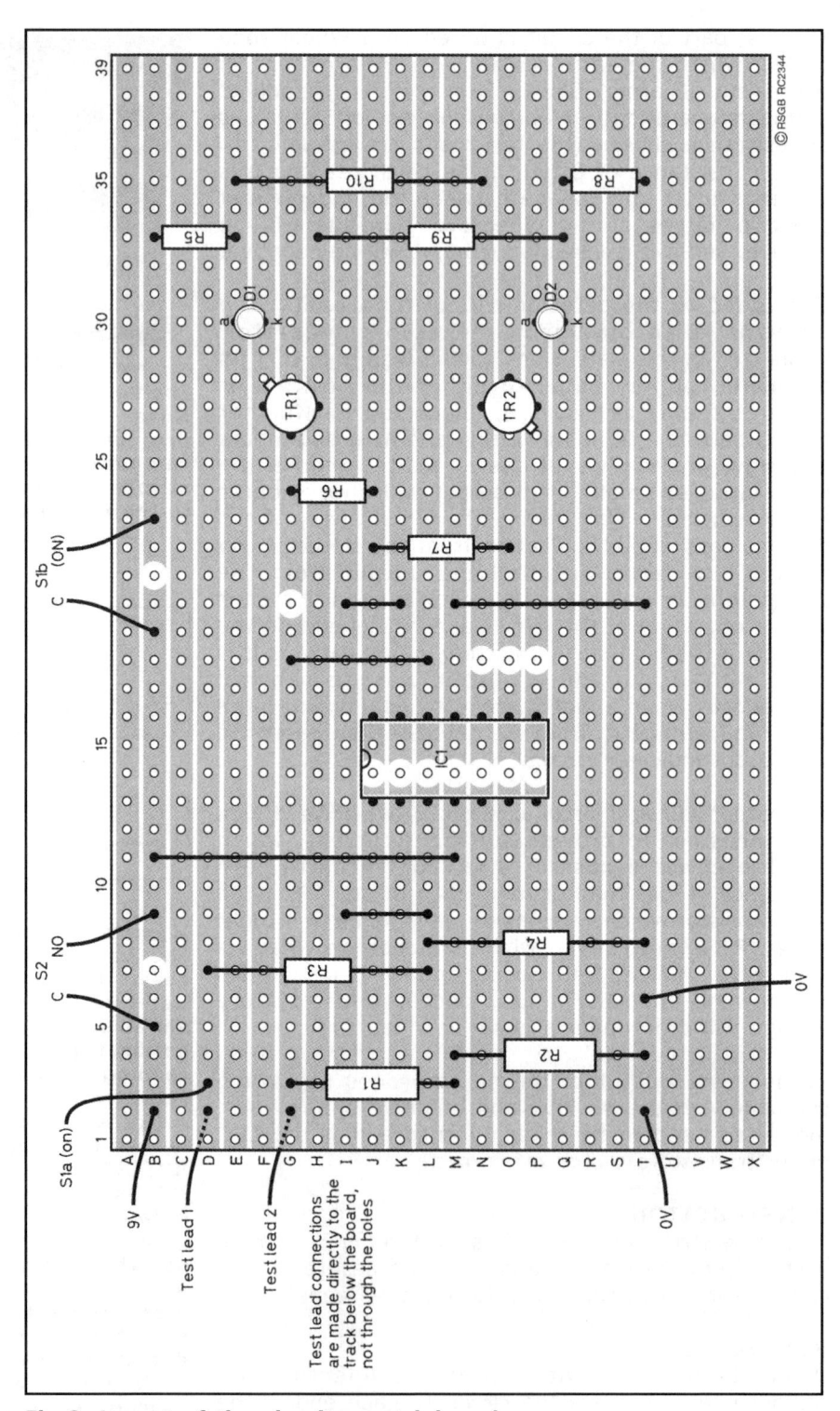

Fig 3. Layout of the circuit on stripboard.

EARTH-CONTINUITY TESTER

mains plug earth pin and any metal part meant to be earthed. Then press the test button. Release the test switch as soon as a pass / fail indicator lights (certainly within 5 to 10 seconds, to lengthen battery life and prevent possible overheating of R1 and R2).

SAFETY NOTICE
The project described here may be used to test the resistance of appliance earth connections, but it is *not* intended to conform to any *legal* requirements for the testing of electrical safety. The RSGB and the author accept no responsibility for any accident or injury caused by its use. *Never* use on mains equipment plugged into the mains – the connection to the mains plug earth pin mentioned in the previous paragraph implies that the plug is free. – Ed

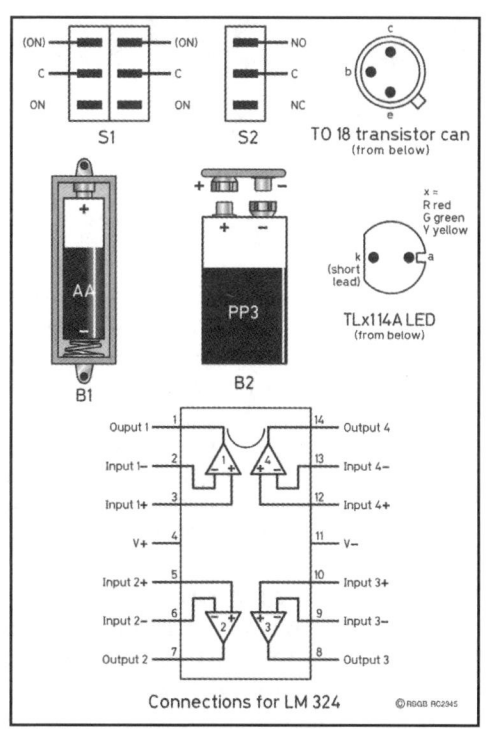

Fig 4. Orientation (and pin-outs) of the batteries, IC1, LEDs and transistors.

PARTS LIST
Resistors - All resistors metal oxide 0.4W 1%, except R1 & R2

R1, 2	0.1, 2.5W
R3, 4	10k
R6, 7	24k
R5, 8	150
R9, 10	330

Semiconductors

IC1	LM324
D1, 2	TLY114A yellow, or TLR114A red and TLG114A green
TR1	BC179 (general-purpose pnp)
TR2	BC109C (general-purpose npn)

Additional items

B1	1 × AA Duracell
B2	PP3
S1	Double pole, momentary on, or push-to-make
S2	SPST

Battery clips / holders
Stripboard
Plastic case*
2 × 4mm plugs & sockets (only required if you are building the project in a case)
2 × crocodile clips

A DIODE/TRANSISTOR TESTER

One of the regular problems when assembling diodes into a project is working out which end is the cathode and which the anode. This is particularly so when using diodes from the 'junk box' and finding that the markings have become somewhat indistinct. It is not too difficult to use a resistance meter to check the polarity but this only looks at one aspect. The unit described below provides an unambiguous means of identifying the cathode and carries out some other simple tests. It can also be used to check that bipolar transistors are 'in the land of the living'.

CAUTION

But first, a few words of caution: be very wary of using rectifiers of unknown characteristics in power supplies or other high-current applications.

The fact that a rectifier looks big enough is not a good indication that it is adequate for the job. Using inadequate devices is likely to result, at best, in poor reliability. At worst, the result could be the well-known 'dark brown smell', followed by smoke and even flames.

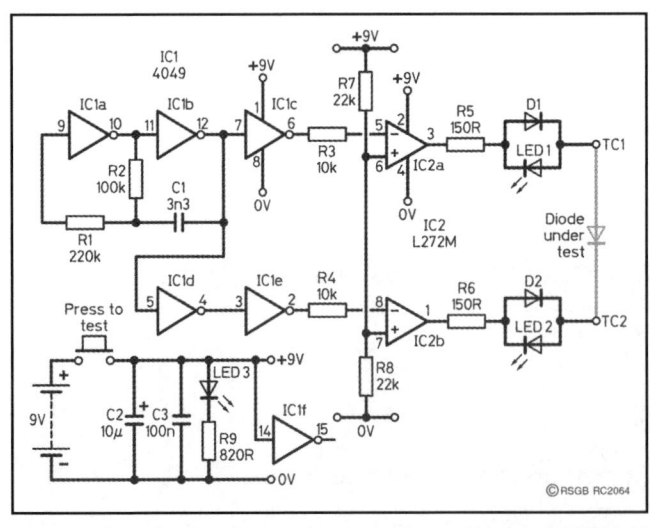

Fig 1. The simple diode tester gives an instant 'dead' or 'alive' indication.

CIRCUIT DESCRIPTION

The circuit is shown in Fig 1. A simple square-wave oscillator is formed by inverters IC1a and IC1b. This runs at about 1.2kHz although the actual frequency is not particularly important.

Gates IC1c and IC1d buffer the output of the oscillator and split the signal into two paths. One of the paths is then inverted, which results in two anti-phase square-wave signals being produced at the outputs of IC1c and IC1e. These anti-phase signals are fed to the inverting inputs of a dual operational power amplifier, IC2, via R3 and R4. The non-inverting inputs are connected to a potential divider chain which holds them at half the supply voltage.

A DIODE/TRANSISTOR TESTER

Consider, now, the diode under test (DUT) being connected to the test terminals, TC1 and TC2, as shown. When IC2a output is high, current will flow through R5, D1, the DUT, LED2 and R6 to the output of IC2b, which will be at about ground potential. The result is that LED2 will illuminate.

During the next half of the cycle, the output of IC2a will be at about ground potential and that of IC2b will be high. Under these conditions, the DUT will be reverse-biased so no current will flow.

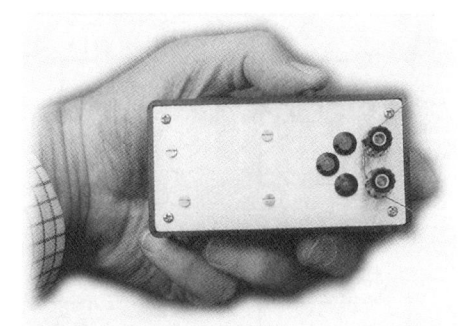

The completed diode / transistor tester.

With the DUT connections reversed, it follows that the operation of the circuit will be similar to that previously described, but LED1 will now be illuminated instead of LED2. If the unit is constructed so that LED1 is adjacent to TP1 and LED2 to TP2, a direct indication is given of the terminal to which the cathode of the DUT is connected.

CONSTRUCTION
With the exception of the placement of C2 and C3, there is nothing at all fussy about the construction and the builder can use PCB, stripboard or 'dead bug' construction as preferred.

Whatever form of construction is used, C2 needs to be connected as close as possible to pins 2 and 4 of IC2, and C3 to pins 1 and 8 of IC1.

A suitable stripboard layout is given in Fig 2. Some care is needed if the SB1A box is to be used. The width of the stripboard is such that it fits between the lid support pillars of the box with little room to spare. Check the placement of the board before drilling the mounting holes.

USING THE TESTER
Connect the diode under test between TP1 and TP2 and press the test button. You have a working device if either LED1 or LED2 illuminates, the illuminated LED

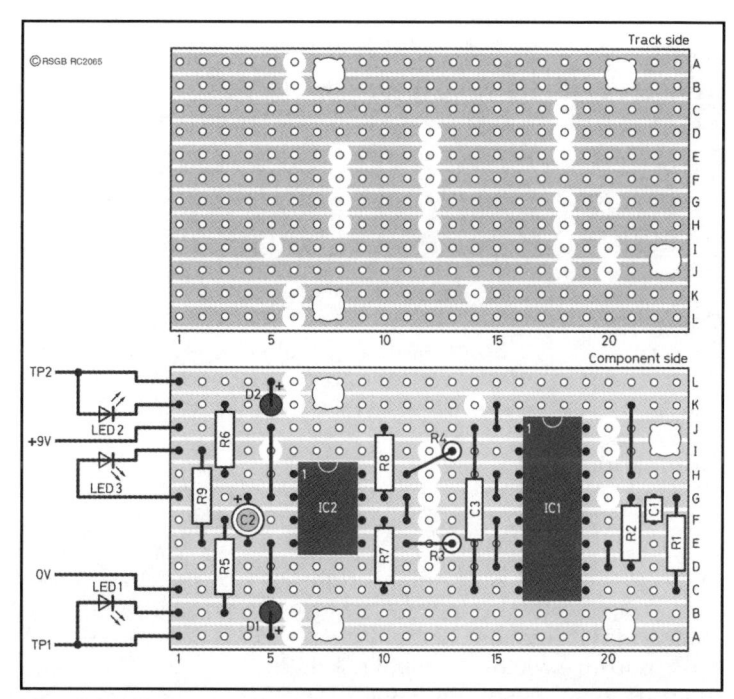

Fig 2. Stripboard layout and component overlay. Note that if you use the recommended project box, there is little room to spare.

85

WEEKEND PROJECTS

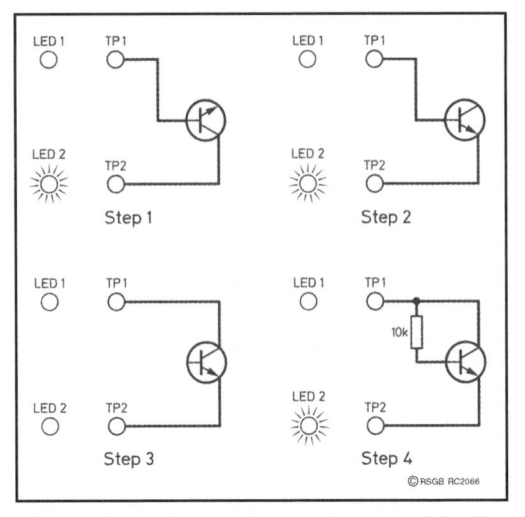

Fig 3. Use of the device to test bipolar transistors.

indicating the cathode of the device. Should both LEDs illuminate at equal brilliance, the diode is short-circuit or is a low-voltage Zener. If both LEDs illuminate but one is brighter than the other, the chances are that the device being tested is a Zener with a breakdown voltage in the 4.7 – 6.8V region. In this case, the brighter of the two LEDs indicates the cathode.

If neither of the LEDs illuminates, this indicates either an open-circuit diode or a flat battery – check that LED3 illuminates to eliminate the latter possibility.

Testing transistors takes a few more steps – see Fig 3. Steps 1 and 2 can be skipped initially and carried out if needed to give information on the probable fault should step 4 fail.

At step 1 it might be found that both LED1 and LED2 illuminate, with one being brighter than the other. This is indicating that the reverse base-emitter breakdown voltage is less than about 7V, which is normal for some transistors. In this case the brighter of the two LEDs is the one to note. The transistor test is not particularly exhaustive but it can be a simple way of checking that a device is still 'alive'.

PARTS LIST
Resistors - All resistors metal film ¼W, 5%
R1   220k
R2   100k
R3, 4   10k
R5, 6   150
R7, 8   22k
R9   820

Capacitors
C1   3.3n ceramic (eg Maplin RA41U)
C2   10μ, 16V tantalum bead (eg Maplin WW68Y)
C3   100n polyester (eg Maplin BX76H)

Semiconductors
IC1   4049
IC2   L272M
D1, 2   1N4001
LED1   3 5mm high-brightness (eg Maplin WL84F)

Additional items
Push button switch, eg Maplin FH59P
4mm terminal posts, eg Maplin FD69A
Project box, eg SB1A – Maplin BZ27E
Stripboard

EXTENDING THE USE OF YOUR DIP OSCILLATOR

The grid dip oscillator (GDO) offers a quick and easy means of checking (to a degree of accuracy acceptable for experimental purposes) the inductance value of coils in the microhenry (µH) range and capacitors in the picofarad (pF) range, such as are commonly used in radio circuits. This can be very useful, for example, when constructing an ATU, a crystal set, a short-wave receiver, a VFO or a band-pass filter for a direct-conversion receiver.

DETERMINATION OF L AND/OR C

For this purpose, I keep with my GDO two fixed-value RF coils of known inductance – 4.7µH and 10µH – and one capacitor each of 47pF and 100pF (but the choice of values is yours). You may decide to keep one or more of each, to be selected from Table 1 and Table 2.

My personal choice of coil type is the moulded RF choke (Maplin) or RF inductor (Mainline or RS). These are axial-lead, ferrite based, encapsulated, easy to handle, and readily available at low cost in a range of fixed values. The capacitors are 5% tolerance polystyrene, also axial-lead.

To determine or verify the value of either an RF coil or a capacitor, simply connect the unknown component in parallel with the appropriate known component to form a parallel LC tuned circuit, ie an unknown L in parallel with a known C (or vice versa), then use the GDO to determine the resonant frequency of the parallel LC circuit.

The value of the unknown component can then be obtained easily to an acceptable approximation, by using the relevant formula from Tables 1 and 2 and a pocket calculator.

Note that, in Tables 1 and 2, *F is the frequency in megahertz as given by the GDO.*

Example 1

An unknown capacitor in parallel with my known 10µH inductor produces a dip at 6.1MHz, hence $F = 6.1$MHz. From Table 1, the value of the unknown capacitor is given by:

$C = 2533 \div F^2$
$= 2533 \div 6.1 \div 6.1$
$= 68$pF.

WEEKEND PROJECTS

Example 2
An unknown coil in parallel with my known 47pF capacitor produces a dip at 12.8MHz, hence $F = 12.8$MHz. From Table 2, the value of the unknown inductance is given by:

$L = 539 \div F^2$
$= 539 \div 12.8 \div 12.8$
$= 3.3\mu H.$

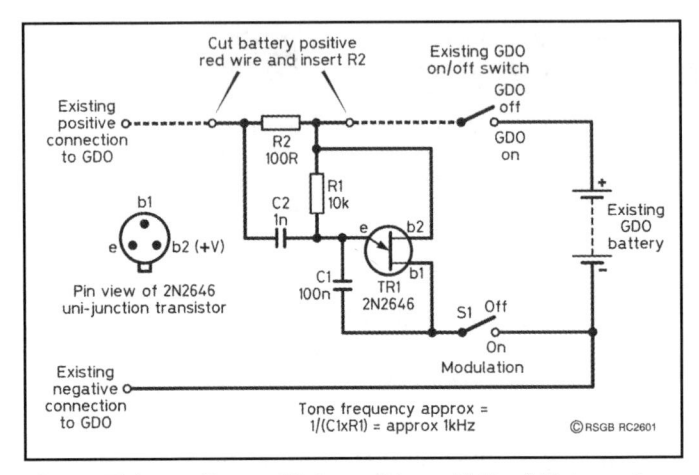

Fig 1. This audio oscillator adds a 1kHz AM tone to a GDO signal.

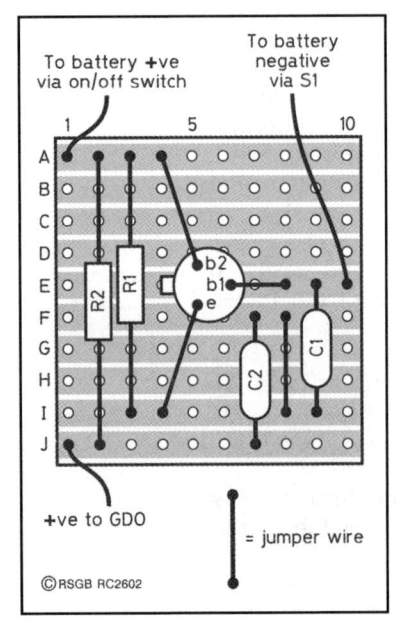

Fig 2. Stripboard layout.

Bear in mind that, because the accuracy of results relies upon the frequency as derived from the GDO, it would be sensible to keep the coupling between the LC circuit and the GDO as loose as possible, consistent with an observable dip. This minimises pulling of the GDO frequency. Also, rather than relying upon the frequency calibration of the GDO itself, it might be useful to monitor the GDO frequency on an HF receiver or a digital frequency meter.

A final point worth considering is that each fixed-value inductor of the type mentioned might have its own self-resonant frequency, but these would typically lie above the HF range so should not be a problem. For example, the self-resonance of my own 10µH inductor is about 50MHz and that of the 4.7µH one is about 70MHz. You could quickly and simply find out the selfresonant frequency of an inductor by taping it to each of the GDO coils in turn and tuning across the full frequency span.

It is best to make L and C measurements at frequencies much lower than the self-resonant frequency of your chosen test-inductor, but perhaps better be safe than sorry and stick with the lower microhenry values if your interest lies between 1.8 and 30MHz.

TONE MODULATION
Sometimes it is useful to be able to hear an audio tone when using the GDO as an RF signal source in association with a radio receiver.

If your GDO does not have tone modulation, you might like to construct the simple add-on 1kHz audio oscillator circuit shown in Fig 1. It uses

EXTENDING THE USE OF YOUR DIP OSCILLATOR

a unijunction transistor, the frequency of oscillation being given approximately by $1/(R1 \times C1)$. The 1kHz tone output connects via C2 to the positive supply line of the GDO, which it modulates.

R2 acts as the modulator load and its value helps to determine the level of modulation. This produces simple but effective tone modulation of the GDO's RF signal which can be heard on an AM or an FM receiver.

Fig 2 gives a suggested layout of the components on a small piece of stripboard, without the need for any track cutting. The finished board might conveniently mount on one of the GDO meter terminals, provided care is taken to isolate the copper-tracks from the terminal.

For known L (µH) of	1	2.2	4.7	6.8	10	22
Unknown C (pF)	$25330/F^2$	$11513/F^2$	$5389/F^2$	$3725/F^2$	$2533/F^2$	$1151/F^2$

Table 1. To determine an unknown capacitance.

For known C (pF) of	10	22	33	47	68	100
Unknown L (µH)	$2533/F^2$	$1151/F^2$	$768/F^2$	$539/F^2$	$373/F^2$	$253/F^2$

Table 2. To determine an unknown inductance.

PARTS LIST
Resistors
R1   10k
R2   100R

Capacitors
C1   100n
C2   1n

Semiconductor
TR1   2N2646

DUAL-VOLTAGE SUPPLY FROM ONE BATTERY

Many projects require a positive and a negative supply voltage, which can of course be provided by using two batteries or a dual power supply. A neater (and cheaper) method, however, is to generate both voltages from a single battery. One way of doing this is to use a simple voltage divider of two resistors but, unfortunately, this method is only really satisfactory if the circuit doesn't consume any significant current. This design allows current to be drawn without significantly affecting the supply voltage.

THE POTENTIAL DIVIDER

Fig 1(a) illustrates the principle of the potential divider. On the left are two 5kΩ resistors connected in series across a 10V power supply. There are no prizes for saying that the voltage at their junction is 5V. If we were to ground this middle point, we would have a dual power supply of ±5V. *This only works, however, if neither side of the 10V supply is grounded*.

To call this a dual power supply is a little misleading. Even though a multimeter connected across each resistor would read 5V, we cannot draw any power from the circuit without drastically affecting its output voltages.

Here is an example. In Fig 1(a), a load of 500Ω is placed across one half of the supply. The parallel combination of the 5000Ω and 500Ω resistors has an effective value of about 455Ω. This can now be considered to be in series with the other 5000Ω resistor across the 10V supply. This now draws, not 1mA as originally, but 1.8mA. (You can verify these figures for yourselves.) We now need to ask ourselves what is the voltage across the 500Ω load resistor? Well, the current is 1.8mA and the resistance is 500Ω, so the voltage is just over 0.9V. So you can see that this simple system cannot deliver any power into an external circuit. Because the original power supply is still delivering 10V, you can see that the negative supply (after connection of our load to the positive half) will now be 10V - 0.9V = 9.1V. Not only has the positive voltage virtually collapsed but the negative voltage has shot up to almost 10V! This system will not fulfil our purposes.

The situation can be improved somewhat if the potential divider is made from lower resistances, as in Fig 1(b). You may like to work out the corresponding voltages here, with and without the load. You will see that the voltage drop on connection of the load is smaller, but look at the current that flows down the divider chain – 100mA – even when there is no load!

DUAL-VOLTAGE SUPPLY FROM ONE BATTERY

This is a very inefficient system which wastes battery energy and of course shortens its life, and also needs resistors capable of dissipating the significant heat generated by the wasted power. If the circuit presents any significant current drain, for example to drive an LED or an audio amplifier, the supply voltage will drop significantly in one or both supply rails.

AN ELEGANT SOLUTION

An effective solution to this problem is to use the potential divider to generate a reference voltage at half the battery supply voltage, but to incorporate another circuit element that prevents the divider supplying any power. This extra element is an operational amplifier (op-amp) connected as a voltage follower, to provide the required load current at this voltage. In this way, the power for the load comes directly from the main power source through the op-amp, not from the potential divider.

As before, we can use the output of the op-amp as our zero voltage level, and the two connections from the original power supply now become the positive and negative rails. This is shown in Fig 2. A battery is ideal as the power source because, as we noted above, neither side of it must be grounded if a true dual-voltage supply is required.

CIRCUIT DESCRIPTION

An op-amp connected this way is using 100% negative feedback, which means that all of the output voltage is fed back to the inverting input. This gives a voltage gain of unity, and means that the output voltage is always the same as the input voltage (hence the name '*voltage follower*').

However, although it is called a 'voltage follower' it is actually a current-controlled device. A change in voltage at the op-amp output caused by a change in load current also appears at its input due to the feedback. This causes the op-amp to change its output current in such a way as to counteract the voltage change on the output, and keep it fixed at the reference voltage. This change occurs effectively instantaneously, and means the circuit responds to all changes in load current demand while at the same time keeping the supply voltages fixed.

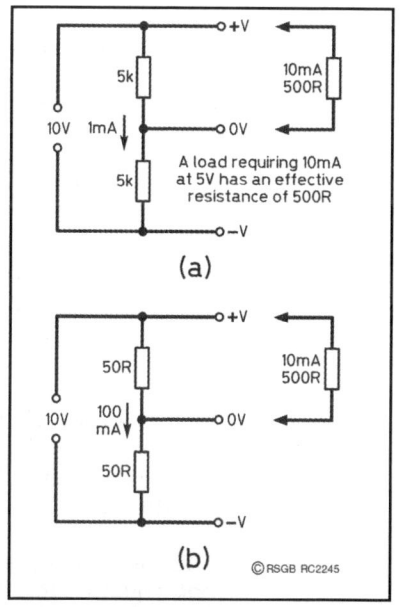

Fig 1. The effect of a load on a voltage divider. (a) A voltage divider of 5kΩ resistors adequately provides V+ and V- of ±5V when there is no load, but if an effective load of 500Ω is connected across the upper resistor it effectively becomes 5kΩ in parallel with 500Ω, ie around 450Ω, and V+ drops to less than 0.5V. (b) If 10 times the required load current flows through the potential divider, ie the load presents 500Ω in parallel with 50Ω, the upper resistor is equivalent to about 45Ω and the loading effect is not so great. Even so, V+ will drop to around 4.5V.

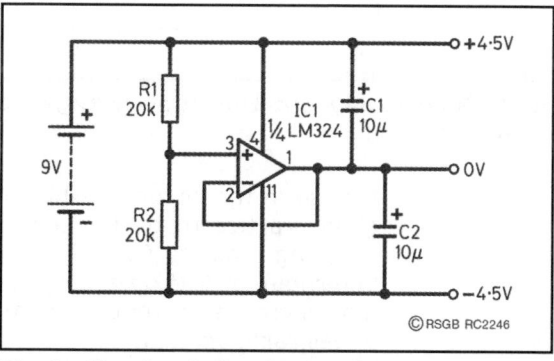

Fig 2. Circuit of the dual-voltage supply.

WEEKEND PROJECTS

VALUES

The component values are not critical, and simple rules-of-thumb are sufficient for choosing values. The main thing to consider is the load current required, which must be within the specification of the op-amp. For example, a typical use of this circuit would be for powering an LM324 quad op-amp. This device will comfortably supply up to 20mA, which should be sufficient for most low-power applications.

The voltage divider resistors should carry around 10 times the input current for the op-amp, which is in the order of microamps, so any value between 10kΩ and 100kΩ will be fine. They do not need a greater power rating than that usually found in low-power circuits, so a 0.4W rating is more than sufficient.

The accuracy of the resistor values determines how closely the values of the plus and minus supply rails match, ie how close the reference voltage is to exactly half of the battery voltage. 1% tolerance is certainly adequate.

The capacitors are smoothing capacitors so, again, values are not critical, and could be increased to 100µF or even 1000µF without much increase in the physical size of the components. The voltage rating for the capacitors should naturally be higher than the output voltage.

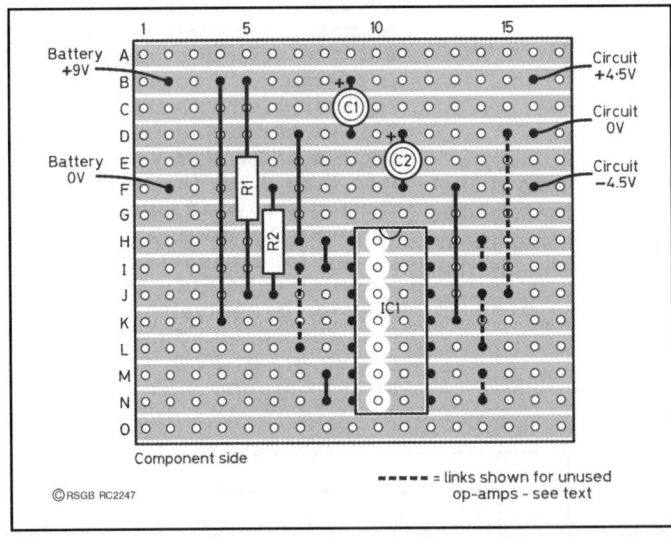

Fig 3. How to lay out the dual-voltage supply on stripboard.

STRIPBOARD LAYOUT

The circuit could be built as a 'stand-alone' project if required and used with different projects as needed. The component count is so low, however, that it's probably just as easy to build the power supply as part of a larger project.

The stripboard layout is shown in Fig 3 for an LM324 quad op-amp, one quarter of which generates the plus, zero and minus supply rails which run along the full length of the stripboard for easy linking to other components.

The other three are available for use in the circuit of which it is part but, if these are not used, it is good practice to tie the output to the inverting input, and the non-inverting input to 0V. This reduces the possibility of any unforeseen behaviour of the op-amps such as latching or self-oscillation, which may increase current consumption or have other unpredictable and undesirable effects.

DUAL-VOLTAGE SUPPLY FROM ONE BATTERY

CONSTRUCTION
Construction is straightforward, the only things to watch out for being the correct orientation of the integrated circuit (check for the position of pin 1 – see Fig 4), and the correct polarity of the electrolytic capacitors (see the markings on their casings).

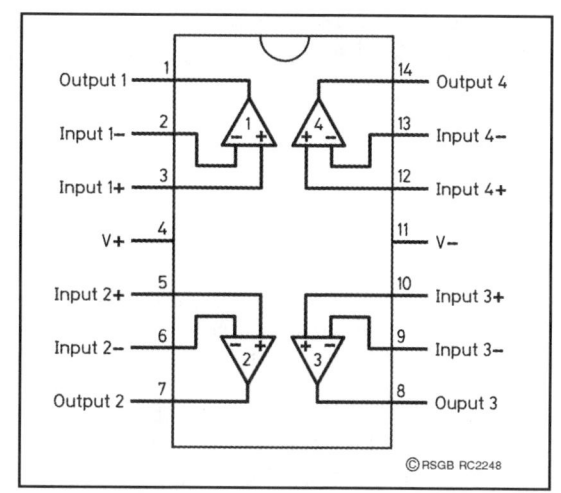

Fig 4. Orientation and internal layout of the LM324 quad op-amp integrated circuit.

PARTS LIST
Resistors
R1, 2   20k, 0.4W, 1%

Capacitors
C1, 2   10μ (see text) electrolytic, voltage rating to exceed the supply voltage

Semiconductor
IC1   LM324 quad op-amp

Additional items
Stripboard
Battery clip

A SIMPLE ELECTRONIC KEYER

Fifty years ago, a Danish amateur, OZ7BO, designed a clever and simple electronic keyer which became known as the *El-Bug*. This device used a double triode valve and a few small components. Some 30 years later the design was updated by G3JIS using two transistors in place of the valve. Both designs used two relays, one of which is used to key the valve transmitter, which had a high positive voltage on the keying line.

These days, most transceivers have keying voltages of less than 12V positive to ground and the keying currents are low. It would be best to check your keying line prior to starting this keyer and, if it is a negative keying line or uses high current, a second relay could be fitted to the keyer output to key the transmitter.

The low power requirement of modern sets enables us to simplify the G3JIS version to 12 components, plus the bonus of a cheap and easily obtainable relay. If all the parts are purchased new the total cost will be about £3.

The El-Bug will not send automatic CQs or make the coffee but, handled properly, it will send perfect Morse at any reasonable speed after a little practice. It can be used with either a single- or double-lever paddle, but will not be *iambic* (ie it does not send alternating dots and dashes if both paddles are held together) and can be built on a 5cm × 6cm board without overcrowding.

CONSTRUCTION

I use an assembly method which is a cross between the sawcut technique advocated by G3VTS ('TT', *RadCom* April 1995) and 'deadbug' construction favoured by G3ROO. All parts are mounted on the copper side of the board and the soldering pads are made large which simplifies construction. This allows for modification to be made with ease. To make a board like this is very easy indeed as no advance planning is necessary providing that more pads are made than you think will be needed.

Start by sticking PVC tape across the board in parallel lines, with 2 or 3mm gaps between them. Next, remove the unwanted tape with a sharp knife and the board is ready for etching in the normal way. No holes are required, as parts like relays and ICs can be stuck to the board 'dead-bug' fashion. The result will be as neat or ugly as you wish but I have built an SSB transceiver using eight boards constructed

A SIMPLE ELECTRONIC KEYER

by this method and they have proved to be invaluable when modifications have been necessary.

At first sight it seems hard to comprehend how such a simple circuit, shown in Fig 1, can generate dots and dashes of correct length and spacing without a binary digit in sight. To understand the operation, first note that the relay contacts used are in the normally-closed position. When the paddle makes contact with the dash side, the supply voltage is applied to C1 which charges, applying a positive voltage to both transistor bases. TR1 saturates, energising the relay and opening the contacts. At the same time, TR2 saturates and keys the transmitter. With the supply to C1 removed by the relay contacts, it discharges via the complex network connected across it. As the base voltage of TR2 drops below 0.6V, TR2 turns off and the transmitter is unkeyed.

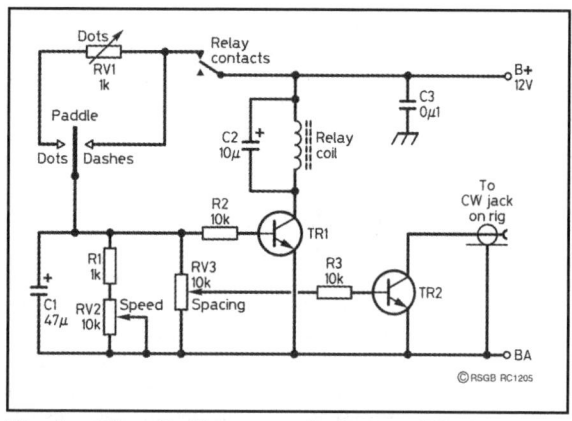

Fig 1. Circuit diagram of the modified OZ7BO autokeyer.

At this moment, TR1 is still on due to its base voltage being slightly higher than that of TR2. As C1 continues its discharge, TR1 turns off and the relay closes again ready for the next operation. If the paddle is kept pressed a stream of dashes will be sent until the paddle is released. The difference in spacing is set by the time difference in the off states of TR1 and TR2. Due to the setting of RV3, TR2 is always switched off slightly earlier than TR1. A similar cycle takes place on the dot side, except that the quantity of charge in C1 is restricted by RV1, thus shortening the dashes to dots.

Correct operation of the El-Bug is very dependent on the characteristics of the relay. In fact it is the mechanical sluggishness of the relay that enables the circuit to operate at all. A 12V relay with a coil resistance of 400Ω is ideal. I use a miniature type YX94C from Maplin. This one is a little fast in operation, so I have retarded it with C2 connected across the coil. The value of C2 with the Maplin relay is not too critical, but a junk box relay may need some experimentation to get correct operation and a slower relay may not need any capacitance. The values of C1 and the three variable resistors are critical; do not leave out R1 or you may short out the power supply!

For initial testing, set the speed pot RV2 at half-track, which should correspond to about 12WPM. With an ohmmeter, set RV3 for 3kΩ between slider and ground. Connect a 12V supply, hold the paddle in the dash position and you will hear the relay clicking away quietly. Connect your rig into a dummy load and connect the key to the keying line. On pushing the dash paddle, dashes should be heard on the sidetone; adjust RV3 for optimum mark/space ratio. Now key the dot paddle and adjust RV1 for correct dot mark/space ratio. With the components specified, the keyer should work between 8 and 25WPM.

WEEKEND PROJECTS

To prevent RF problems, the keyer should be built in a metal box and coaxial cable used to connect to the key jack of the transmitter. I had problems on 15m but this was cured by fitting two 10nF ceramic capacitors, one from the collector of TR2 to ground and the other between its base and ground. With these fitted, full-power operation was available on all bands.

OPERATING

A newcomer to auto-keyers may well find first efforts depressing, as the keyer seems to have a mind of its own. I remember 10 years ago borrowing a keyer for a few days and, although the marvel of it was obvious, I was unable to master it in five minutes and gave up. Now the beauty of making your own for a few pounds is the great psychological advantage you gain. Because you made it yourself and because it actually works, you feel obliged to persevere. You may well give up and put it away for a while but, sooner or later, you will go back for another go.

After three and a half hours' practice, I was able to do practice-QSOs at 12WPM and soon ventured on the air. I made mistakes, but slowly got better and the El-Bug has revitalised my interest in Morse and makes sending a pleasure. Paradoxically, sending good Morse has improved my receiving as well!

PARTS LIST
Resistors
R1   1k
R2, 3   10k
RV1   1k preset
RV2   10k linear potentiometer
RV3   10k preset

Capacitors
C1   47µ, 25V
C2   10µ, 25V
C3   0.1µ, 60V

Semiconductors
TR1, 2   BC108, BC109, BC171 or near equivalent

Additional item
Relay   Maplin YX94C

FIELD-STRENGTH MEASUREMENT

Never expect the same antenna to work the same way in two locations. Ground conductivity and reflections from objects, even at considerable distances, can upset our expectations. The effects are more pronounced on VHF, and experimentation with a field-strength indicator (not 'meter', because no absolute accuracy can be claimed) can be extremely enlightening.

USE OF THE INDICATOR
You can get some idea of the field-strength pattern of your antenna by taking a reading on your completed field-strength indicator, then walking around and noting how the reading changes.

CONSTRUCTION
This indicator was originally designed for a specific use: testing glider radio systems. Because the indicator has subsequently proved so useful for all amateur radio bands from 160m to 70cm, it is worth passing on.

Several designs were tried to begin with, the most interesting being a dipole with a 68Ω resistor connected between the two halves and an RF detector connected across the resistor. With the dipole securely mounted on a tripod, the dipole length could be adjusted for maximum reading, enabling the wavelength of the signal to be measured with surprising accuracy.

Unfortunately, the sensitivity of the unit was not sufficient for the purpose, so the resistor and detector were replaced by a full-wave bridge as shown in Fig 1.

A wide-band amplifier could have been added to the original circuit to gain the necessary sensitivity, but experience has taught me that it is far better to use a 'passive' unit than one requiring a power source, as the battery has the annoying habit of being flat when needed most urgently.

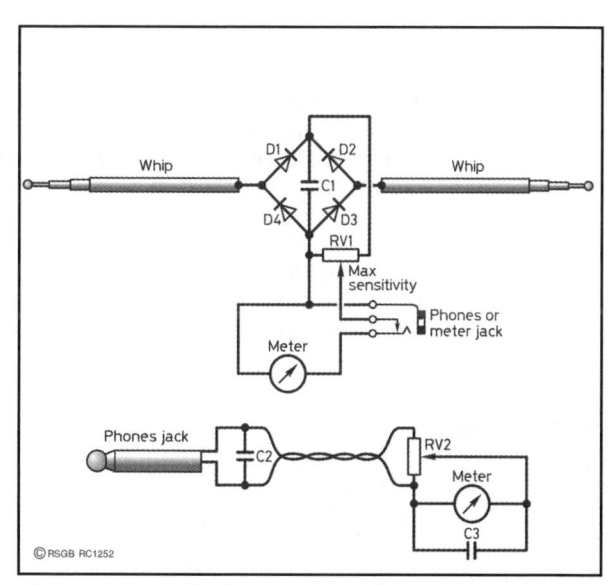

Fig 1. Circuit diagram of the field strength indicator

WEEKEND PROJECTS

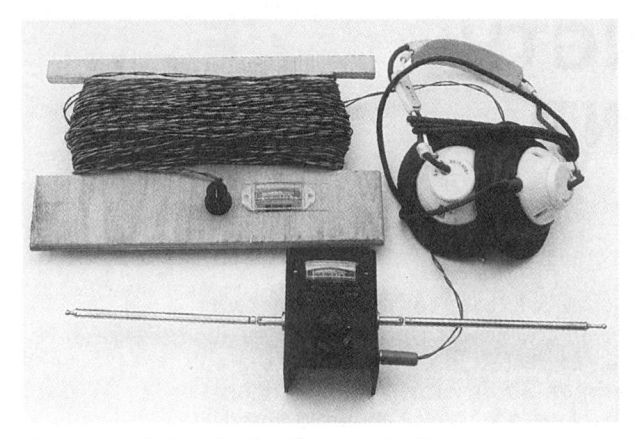

The completed indicator ready for use.

In the circuit diagram there is a jack socket labelled 'phones'. This has two functions. On the aircraft band, amplitude modulation is used, and the modulation can be checked by plugging a pair of high-impedance headphones into this socket. Unfortunately, FM is the most popular mode used on the amateur VHF bands and this is not rendered audible using a diode detector.

The second function enables an extension meter to be used. This consists of a second meter and sensitivity control which, when used, requires the sensitivity control in the head unit to be set to maximum. The extension cable can either be two lengths of hook-up wire twisted together or a length of coax. If you have a length of coax in your junk box which has been removed from service as it has become lossy at RF, it would suit this job admirably because a DC measurements are unaffected by the problems that cause the cable to be lossy at VHF.

A simple way to twist long lengths of hook-up wire is to lay out two lengths of wire on the ground and make the far ends fast to a solid anchor such as a bench vice, tree or fence post. The two free ends are then clamped into the chuck of a hand drill and, with the wires under light tension, the drill is used to twist them together.

The SWR bridge is very useful when tuning matching networks at the base of antennas. but can be misleading at times. Spurious resonance within the matching network can, on occasions, indicate a good VSWR when there is a poor match to the antenna. Using a field-strength indicator on a tripod some distance from the antenna and an extension meter alongside the operator enables quick and easy tuning-up of systems.

There is little to be said about the construction of the indicator, with the exception that all wires should be kept as short as possible. The meter recommended is a 50μA unit. However, a 200μA meter from the junk box was used; this reduced the sensitivity but was very cheap!

```
PARTS LIST
RV1, 2          10k linear potentiometer
C1-3            10n ceramic
D1-4            OA2 germanium diode or similar
Meter           50μA FSD
Antenna halves  560mm telescopic whips
Jack socket with switch and plug to match
```

A FRAME AERIAL FOR HF

A receive-only aerial does not have to be efficient to be effective. The most important attribute for receiving is the signal-to-noise ratio. Because the noise source is usually localised, a directional aerial can often be used to remove noise, leaving the required signal in the clear.

THE PROBLEM
Tuned ferrite rod aerials can be very good at achieving this, but I have always found that on HF they are not as effective as they are on the LF bands. The frame aerial is far superior and you do not have to find suitable ferrite!

THE SOLUTION
To achieve a wide tuning range, the usual solution is to have a large coil of wire and select sections of it using a switch. A variable capacitor is then used to resonate the coil on the band in use.

My solution was to make the loop resonate at several frequencies at the same time. The circuit diagram is shown in Fig 1. C1, C2 and C3 are the three gangs of a 500pF tuning capacitor, as found in older broadcast receivers and at rallies. L1, L2 and L3 are sections of a single-frame coil as shown in Fig 2 and tapped at two points. The circuit's behaviour is very complex, with at least five resonant frequencies for any given setting of the capacitor, but the three main ones cover the amateur bands.

The tuned circuit formed by L3, C2 and C3 is designed to cover the range 11 to 30MHz for a full sweep of the tuning capacitor. L2 with C1 and C2 resonate at about 7 to 15MHz, and L1 with C1, C2 and C3 cover 1.5 to 4MHz. It must be expected that even a slight variation in design will vary these bands, but the differences between my Mk1 and Mk2 efforts were slight, the main difference being that Mk1 worked up to 35MHz, whereas the final one just reaches 30MHz.

Coupling the aerial to the receiver is accomplished by a single-turn loop round the outside of the frame aerial.

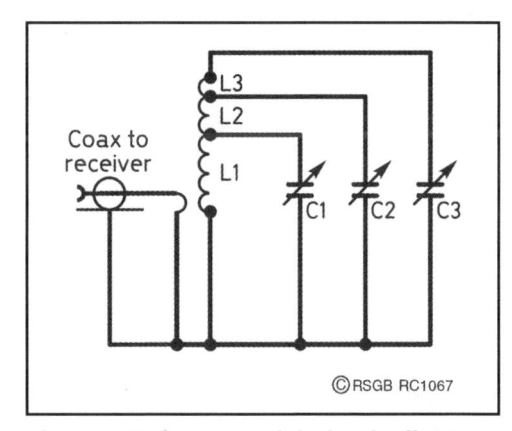

Fig 1. HF frame aerial circuit diagram.

WEEKEND PROJECTS

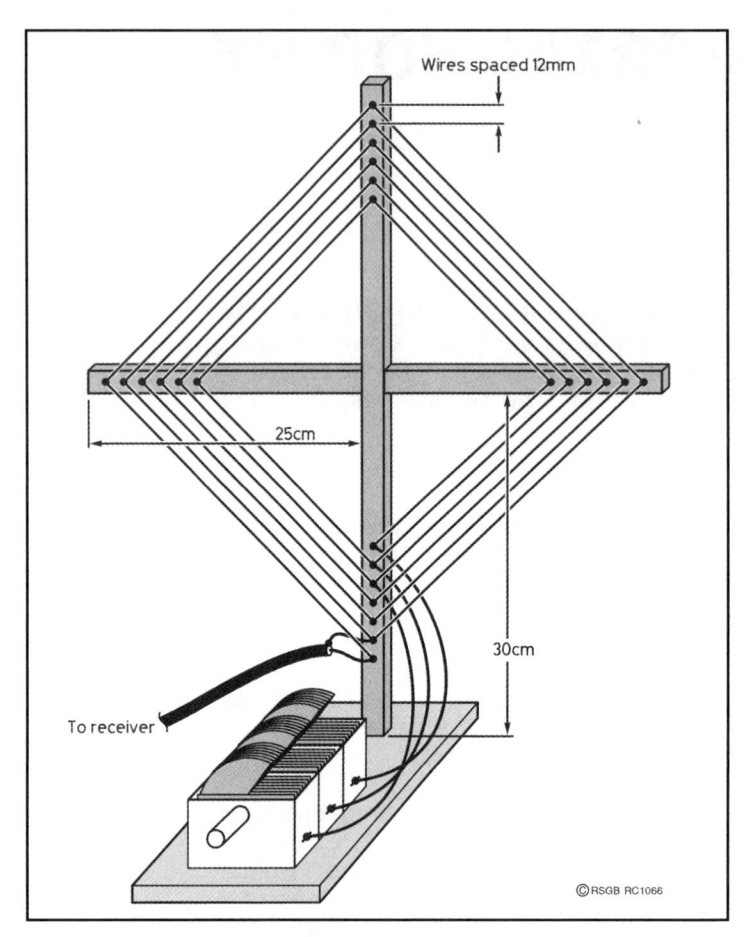

Fig 2. Construction of the frame aerial.

CONSTRUCTION
This is up to the constructor to a large degree, but I used 15mm square timber, half-slotted to make the cross. I then used six 10mm pins hammered in at the ends of three of the arms, leaving 3mm exposed. The fourth and lowest arm needs seven pins. These pins were used to support the tinned copper wire, which was then wound round the pins and soldered to hold it in place, as Fig 2 illustrates. The three-gang capacitor was fixed to the base plate and short lengths of wire used to connect to the frame aerial.

COMPUTER-TO-RADIO INTERFACES

Have you ever wanted to try those digital modes that require connecting your computer sound card to your transceiver, but have been put off by the cost of commercial interfaces? Here are three solutions – a fully-isolated interface, a simpler interface without isolation, and an even simpler one for hand-held radios. Use them for RTTY, PSK31, SSTV, CW, MFSK, FeldHell, EchoLink, eQSO...

Digital communication modes represent one of the fastest-growing areas of interest in amateur radio, with the past decade having seen many developments. Over the past few years, new data modes like MFSK and PSK31 have become popular.

GOING DIGITAL?
For digital transmissions such as RTTY, PSK31, MFSK and others, you need to be able to connect your computer to your transceiver in an effective and consistent way, allowing signal levels between the two to be correctly set. [Much has been said recently about this subject [eg 1, 2], and the reader is encouraged to consult both of these articles if the intention is to build any of these interfaces – *Ed.*]

An interface unit allows transmission and reception of these modes without the expense of purchasing a separate TNC or DSP device. A regular sound card, as found in most of today's computers, can easily handle DSP functions. The interface circuits given here are designed to operate without an external power supply.

INTERFACE CIRCUITS
There are various circuits to enable you to build your own interface. I have included here some simple designs that I have built, tested and which work very well, considering their simplicity and economy. They will also perform well if you intend to run an Internet gateway using *eQSO* or *EchoLink* software. Software for these modes is freely available via the Internet.

Some PTT techniques make use of the transceiver's VOX circuits – but don't forget to disconnect it or the inevitable computer beep or late night MP3 might create a surprise or two! I have avoided requiring VOX switching here, but many new PCs have no RS-232 port, so I think we are soon going to have to find another way to switch our radios.

Digital modes can have a long transmitter duty cycle. Try to keep your output power to 10 – 20 % of the maximum rated power. Disable all the rig compressors, DSP noise reduction etc.

WEEKEND PROJECTS

The fully-isolated interface

Fig 1 incorporates two 600Ω audio transformers (T1, T2) and an RS-232-driven optocoupler, IC1. Preferably, use an IC socket for IC1, for possible quick replacement! All components can be obtained from Mode Components (see list).

The purpose of the transformers and an optocoupler is to isolate the transceiver from the computer, keeping the interference from the PC to a minimum. Ensure that the screening on the radio and the screening on the PC are not connected together.

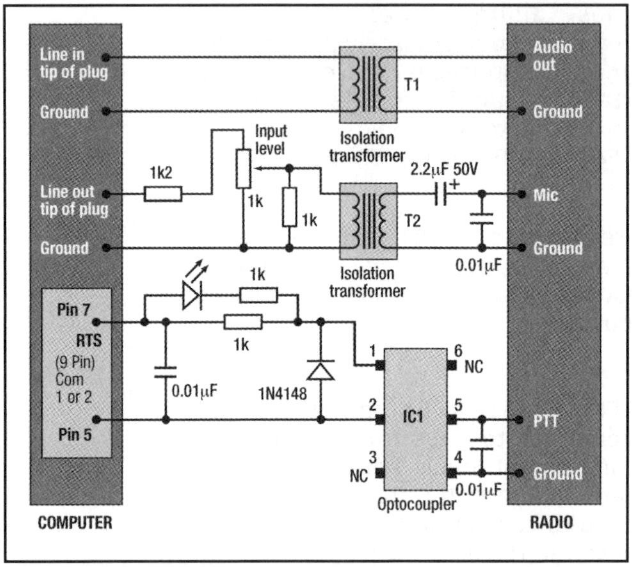

Stereo 3.5mm plugs connect the line in and out on the computer soundcard. Use the tip and earth only because, in this application, the sleeve is not used.

To control the radio PTT, an isolated signal from the computer's RS-232 RTS line is used. If you have an available DB-9 connector on your computer, RTS is pin 7, and ground is pin 5. If you have a DB-25 connector, RTS is pin 4, and ground is pin 7. [Remember to configure your software to use the RTS pin for PTT – *Ed.*]

Fig 1. The fully-isolated interface circuit.

COMPONENTS FOR Fig 1
Resistors:
3 x 1kΩ
1 x 1.2kΩ 0.25W
1 x 1kΩ linear pot

Capacitors:
1 x 2.2µF 50V
3 x 0.01µF

Transformers: 2 x 600Ω, type 9000 (RS 208-822)
Integrated circuit: 1 x IC1 optocoupler 4N25 (RS 597-289)
Miscellaneous: 1 x red LED (high-sensitivity type)
1 x diode 1N4148
2 x 3.5mm stereo jack plugs
1 x 9-pin D-plug (for RS-232 port) & cover
Screened cable, project box

COMPUTER-TO-RADIO INTERFACES

To control the audio going to the microphone input on the transceiver, a 1kΩ potentiometer varies the input to T2, and is adjusted for correct audio drive to the radio, converting line signals (0.5V) to microphone (10mV) levels. The value of the 1.2kΩ resistor (connected to 'line out') can be changed if you are troubled by the pot always being at the bottom or top of the range or, alternatively, by adjusting the computer's 'audio out' slider [1] until the correct level is achieved.

Operationally, audio levels are adjusted by the computer level controls or may be incorporated in the software you will be using.

The LED (high-sensitivity type) is used as an indicator when the interface is in the transmit mode.

It is suggested that the finished interface is put in a metal box and that the grounding is taken from the *radio* side of the circuit.

The simple interface

This circuit (Fig 2) is very similar to Fig 1 except it does not use audio transformers or the optocoupler, but performs splendidly.

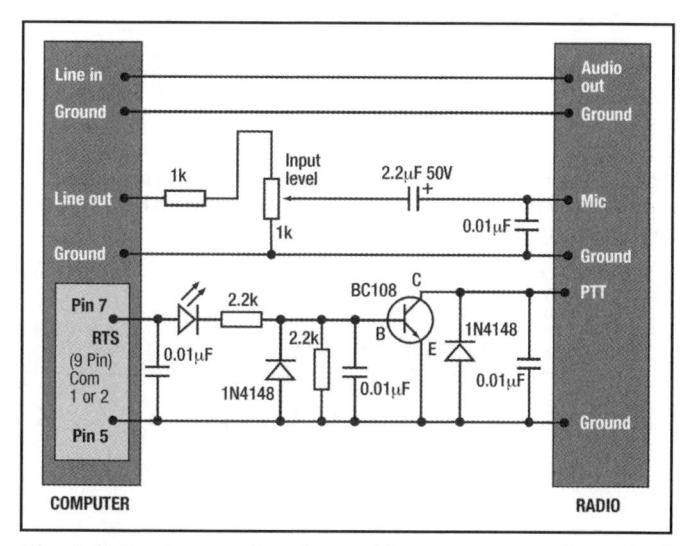

Fig 2. Interface having the same electrical performance as Fig 1, but without isolating transformers and optocoupler.

In this circuit the computer RTS drives an open-collector transistor for the PTT. You can use any general npn transistor instead of a BC108.

COMPONENTS FOR Fig 2
Resistors:
1 x 1kΩ
2 x 2.2Ω 0.25W
1 x 1kΩ linear pot

Capacitors:
1 x 2.2µF 50V
4 x 0.01µF

Miscellaneous:
1 x red LED (high-sensitivity type) - 2 x diode 1N4148
2 x 3.5mm stereo jack plugs
1 x BC108 transistor
1 x 9-pin D-plug (for RS-232 port) & cover
Screened cable, project box

WEEKEND PROJECTS

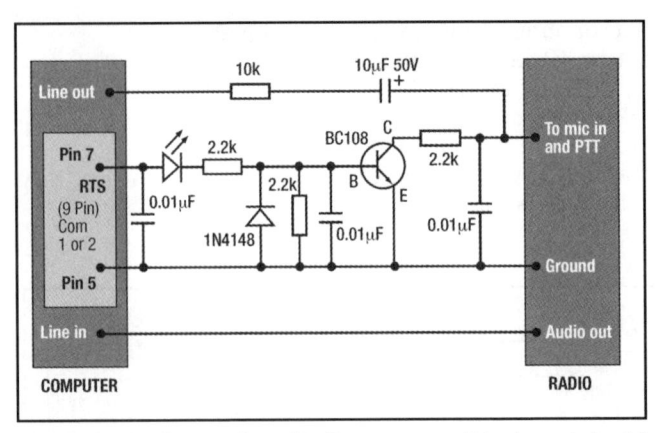

Fig 3. Interface circuit for use with hand-held transceivers where the microphone and PTT connections are joined.

A simple interface for handheld radios

A handheld's microphone and PTT are normally combined, hence the circuit of Fig 3 was designed.

Audio levels can be adjusted only by the computer's level control. Stereo 3.5 mm plugs connect the 'line in' and 'line out' sockets on the computer sound card. Use the tip and earth only as, in this application, the sleeve is not used.

Resistors:
3 x 2.2kΩ
1 x 10kΩ 0.25W

Capacitors:
1 x 10µF 50V
3 x 0.01µF

Miscellaneous:
1 x red LED
1 x diode 1N4148
2 x 3.5mm stereo plugs
1 x BC108 transistor
1 x 9-pin D-plug (for RS-232 port) & cover
Screened cable – project box

REFERENCES
[1] 'In Practice', *RadCom* January 2004, pp59/60.
[2] 'A Useful Audio Level Indicator', R G Dancy, G3JRD, *RadCom*, March 2004, p49.

ON THE WEB
EchoLink   www.echolink.org
eQSO   www.eqso.org
Mode Components   www.modecomponents.co.uk

A CHEAP AND CHEERFUL APPROACH TO KEYING

Here is a design for an electronic keyer, the aims of which were to be battery operation for portability, and no moving parts to avoid the necessity for complicated mechanical construction. This is what emerged, and how its has developed over the years.

SOME HISTORY
Keying levers in the conventional keyer were replaced by touch-sensitive switches which, in turn, drove a multivibrator, or a multivibrator-plus-divider, to produce the dots and dashes. These were positive with respect to 0V and drove an open-collector npn transistor which keyed the transmitter. All the ICs were CMOS, so the battery consumption was negligible. Now, 20 years later, the keyer is only on its third battery (a 9V PP3)! The touch-sensitive switches were aluminium pads screwed on to the top surface of the keyer, and were 'keyed' by pressing the third finger on the central pad and tapping the Dot and Dash pads with the index and fourth fingers. Because of this keying action, I dubbed it the 'Piano Keyer'. A sidetone oscillator was included, which was useful for off-air practice – very necessary for a new style of keying! Although published at the time it didn't attract much attention and, as far as I know, it remains ignored.

Most of today's transceivers incorporate an integral electronic keyer. This internal keyer requires three connections – Dot, Dash and the Chassis Ground or 0V line. Grounding either the Dot or Dash contact triggers a stream of either dots or dashes. Grounding both Dot and Dash contacts simultaneously triggers a stream of alternate dots and dashes. This is known as 'Iambic Keying', where the keyer recognises that both the Dot and Dash contacts are grounded, as opposed to earlier keyers that recognised either the Dot or Dash contact, and where (possibly) the Dash contact would override the Dot contact.

The convenience of having an internal keyer in one's transceiver has made the conventional stand-alone keyer, 'bug' key or Vibroplex, and the Piano Keyer, obsolete. However, one still needs a device to make the contact between Ground and Dot and Dash contacts and this is usually achieved by a 'paddle'

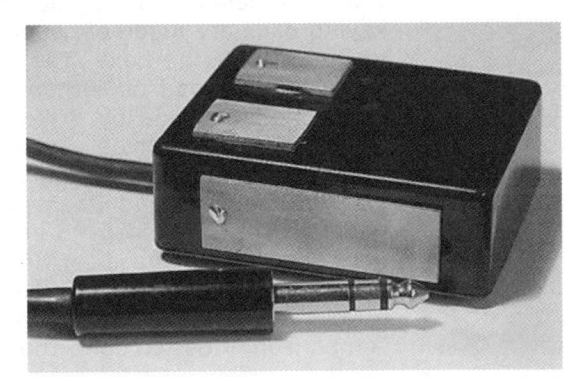

The new Piano Keyer, showing the dot pad on the top left and the dash pad on the top right of the box, with the earth pad on the left-hand side of the box. There is an identical pad on the opposite side.

WEEKEND PROJECTS

keyer which can be likened to a pair of manual Morse keys ('pumphandle' keys) mounted back-to-back and operated with sideways movements of the levers. In practice, the paddle keyer is mechanically quite complicated, as both levers have adjustable contact gaps and spring tensions, and the whole is mounted on a heavy base-block to prevent it walking across the table under the influence of the sideways movement of the keying levers. The better models are beautifully engineered but, in view of the very simple function that they perform, are horrendously expensive. The paddle keyer, being mechanically complicated, is possibly beyond the capabilities of the average small electronic workshop. With this in mind, it seemed a good idea to have another look at the Piano Keyer.

THE NEW PIANO KEYER

For the new Piano Keyer, all we need from the original are the touch-sensitive switches, as the production of dots and dashes is now performed within the transceiver. The circuit is now reduced to one IC, a 4011 (a CMOS quad two-input NAND gate), and is shown in Fig 1. In its quiescent state, the output on both the dot and dash lines is high, eg +9V. When the 0V and Dot pad are bridged by finger contact, pin 9 goes towards 0V and the output of IC1c on pin 10 goes to +9V. Both inputs of IC1b, pins 5 and 6, are now +9V, and the output on pin 4 goes to 0V, thus enabling the transmitter's keyer. Operation of the Dash keyer is identical.

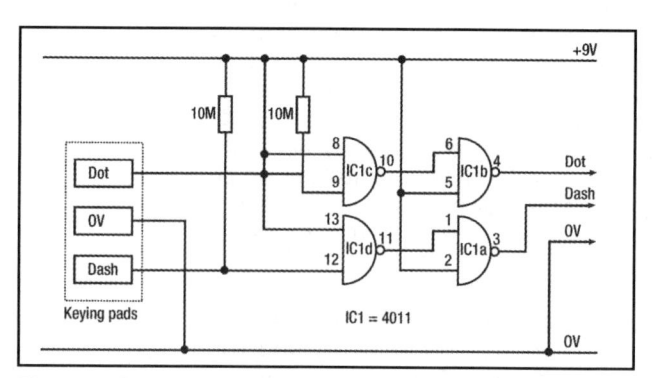

Fig 1. Circuit diagram of the 'Piano Keyer'.

The new Piano Keyer is built into a small Maplin project box of 75 x 55 x 25mm, which is just large enough to accommodate the 9V battery and small circuit board which are held in place by double-sided sticky pads. The Dot and Dash keying pads are made of copper-clad circuit board and are glued on to the outside of the box. The Dot and Dash pads are on the top and there is a 0V pad on either side, (see the photograph). Electrical connection is made by drilling right through the pad and box and soldering on the outside. In operation, the keyer is held on either side between the thumb and fourth finger [just like a computer mouse – Ed.] and the Dot and Dash pads are 'keyed' by the index and third fingers. This may seem to be unusual at first, but I believe that it is not difficult to get used to. If you are reading this, you have probably graduated from a 'pumphandle' key to a 'bug' or Vibroplex and thence to an iambic keyer. The cultural leap to the Piano Keyer is minimal in comparison – easier done than said!

I am indebted to John, F5VJH (ex MM0BPO), who suggested the new form of construction because he felt it would be easier to operate than the original Piano Keyer which has the three pads in line on top of the box. Either way, one unforeseen advantage is that the keyer has no

A CHEAP AND CHEERFUL APPROACH TO KEYING

tendency to 'walk' across the operating table! With such a simple circuit, the operator can choose the physical layout which suits him/her best, and even the conventional vertical paddles can be closely replicated.

This circuit depends on reasonable skin conductivity for reliable operation. For the operator who has exceptionally dry skin, an occasional lick of the fingers may be necessary! Of course, microswitches could be configured to do the same job, but then we're back to mechanics...

A USEFUL AUDIO LEVEL INDICATOR

> Many amateurs, when wanting to try the many digital modes on offer, find that connections must be made from their transceivers to their computers. In the January 2004 *RadCom*, Ian White, G3SEK, explained how to make these connections, and gave advice on how to set the transmit audio levels so that overdriving and non-linearity were avoided. Here is an excellent piece of ancillary equipment to enable you to monitor the levels you set, thus ensuring complete repeatability when setting up.

In the January 2004 *RadCom*, G3SEK, in his 'In Practice' column, described the problems of setting transmit audio levels for PSK31 operation; a similar problem exists on receive for the NOAA and other weather satellites requiring software for the computer sound card. Several years ago, a useful level indicator was designed and built, which has been in use ever since, for several modes including PSK31 and *WXsat*.

WHAT DOES IT DO?
Connection of the indicator to the input terminals of the final piece of equipment in the chain enables the level for correct operation to be achieved easily. Receive calibration should be by the traditional hit-and-miss method, reviewing the results of each test until the optimum is found, the procedure being quite straightforward. Then you can make a note of the indicated level for future use. When using the device to monitor your audio transmit level, particularly for PSK31, use the technique described by G3SEK to set the level, and then use the indicator to register its magnitude, enabling you to set it precisely again when you have been using other modes.

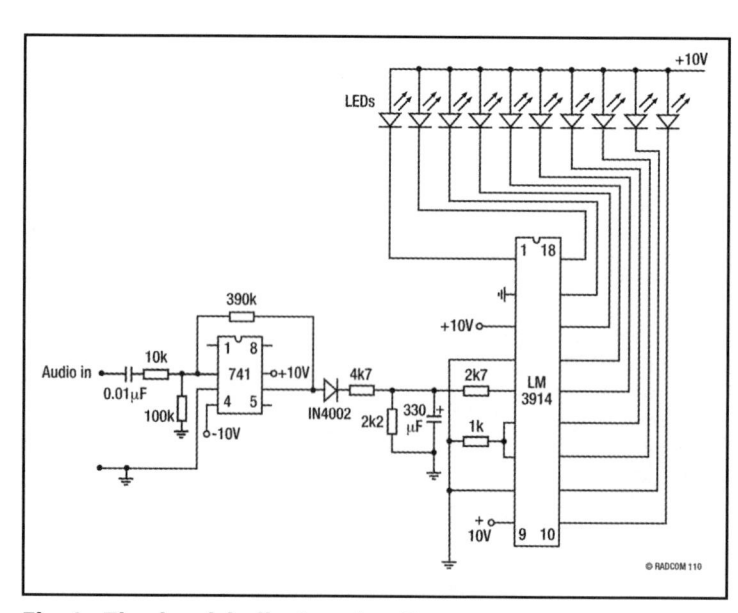

Fig 1. The level indicator circuit.

A USEFUL AUDIO LEVEL INDICATOR

THE CIRCUIT
The original indicator, the circuit of which appears in Fig 1, was built on a piece of stripboard about 5cm x 12cm (2in x 5.75in), and drew its power from the RIG RX2 receiver, but a simple power supply utilising a 12-0-12V transformer, two diodes and an electrolytic capacitor could be used.

Note that the 741 op-amp requires ±10V. This balanced supply is derived partly from the circuit of Fig 2, which provides +10V regulated from a +15V unregulated supply, and Fig 3, which produces -10V from the single +10V supply.

In the circuit, the audio signal is applied to the inverting input of the 741 op-amp, the output of which is rectified and, after integration, an LM3914 LED bargraph display driver is used to provide visual level indication by means of 10 discrete LEDs or a single bargraph module.

Fig 2. Deriving the +10V regulated supply from a +15V unregulated source.

AN ALTERNATIVE
A sensitive moving-coil meter could easily be used here, in place of the LM3914 and LEDs, although it has not been tried. The LED option is probably less expensive and is certainly very robust.

No claim is made that the design is an optimum one, but it worked as soon as it was switched on, and has proved to be completely reliable. The unit could probably be used as a training project for potential radio enthusiasts, as well as being a useful piece of gear around the shack.

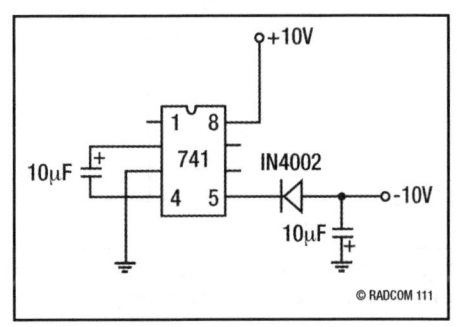

Fig 3. -10V is generated from the +10V supply.

A 'LOOP' ALARM

Putting on a station at a local fair or similar function is a good way of gaining publicity for our wonderful hobby but, when you are on a stand, it is not always possible to keep an eye on the equipment. This alarm is very useful for those situations. It is similar to those used in shops, where a loop of wire is wrapped around the goods. If the wire is broken to remove an item, the alarm sounds.

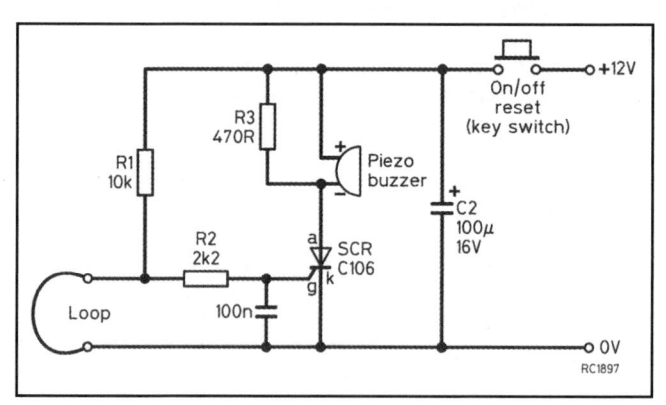

Fig 1. The loop security alarm uses the wire loop to prevent the SCR from being turned on.

HOW IT WORKS

The alarm uses a *thyristor*, which is also known as a *silicon controlled rectifier* (SCR). [To avoid any misconception by reading the name wrongly, this is a *rectifier* that can be *controlled*, and it is made out of *silicon – Ed.*] An SCR is like any other diode in that it will only allow current to flow in one direction, ie from the anode, 'a', to the cathode, 'k'. However, unlike any other diode, this current will only start to flow when a small positive voltage is applied to the gate, 'g'. Having started to conduct, the SCR continues to do so, even if the gate voltage is removed. In the circuit, shown in Fig 1, the SCR does not conduct under normal conditions because the wire loop maintains zero volts on the gate. If the loop is broken, the gate is pulled positive by R1 and R2, which makes the SCR

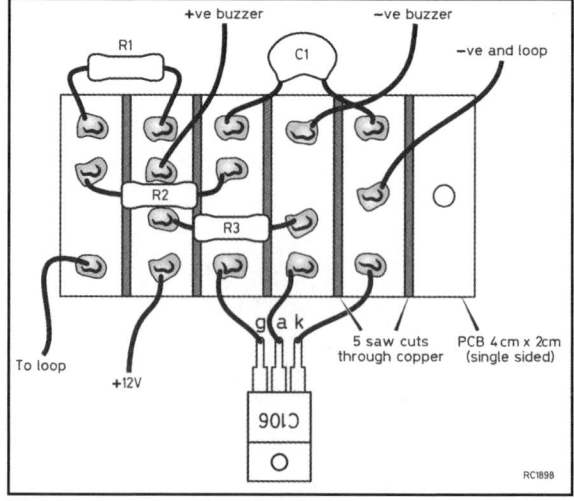

Fig 2. Construction is simple on a small piece of PCB, but be careful when cutting through the copper not to cut right through the board!

A 'LOOP' ALARM

conduct, sounding the alarm. Even if the loop is re-joined, the SCR continues to conduct and the alarm continues to sound until the power is turned off.

CONSTRUCTION
The circuit is built on a small piece of single-sided PCB. Saw cuts are made through the copper, as shown in Fig 2, to form pads. The components are soldered to these pads. The on/off switch should ideally be a key-switch, to prevent the alarm being switched off by an unauthorised person.

The ends of the loop can be connected via practically any type of connector to the case containing the alarm; indeed, the loop could comprise of any number of short lengths of wire, joined with in-line plugs and sockets. The latter would allow a single item of equipment to be removed without disturbing the whole loop.

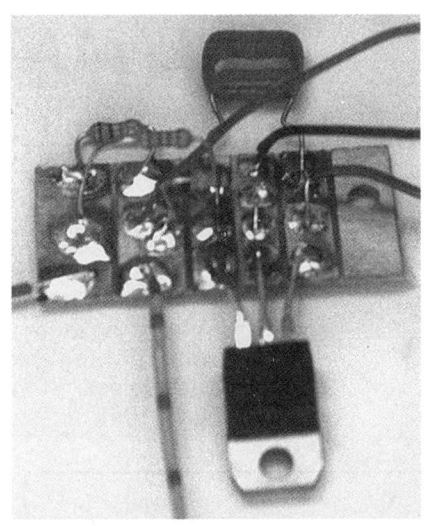

The completed project. In this photo the key-switch cannot be seen, as it is on the far side of the box.

PARTS LIST
Resistors - All resistors $^1/_4$W
R1   10k
R2   2.2k
R3   470

Capacitors
C1   100n
C2   100µ electrolytic

Semiconductor
TH1   C106 (Maplin)

Additional items
Key-switch
Piezo buzzer (Maplin)
Plastic case
PCB material
Wire for loop
Plug(s) and socket(s) for loop

AN L-MATCH ATU

Many aerial tuning units (ATUs) have been tried, but good results are regularly obtained from an L-match tuned against quarter-wave counterpoises (elevated 'ground wires'). Actually, only two counterpoises, 20m and 5m long, are needed to cover all bands from 80m to 10m.

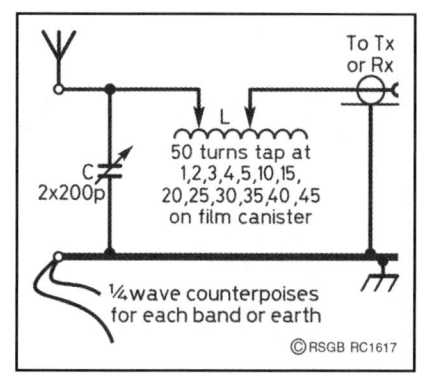

Fig 1. The L-match is constructed from just two components.

CONSTRUCTION

The L-match employs just two main components, a coil and a capacitor, as shown in Fig 1. The coil is wound on a plastic 35mm film container. It has a total of 50 turns of 24SWG enamelled copper wire, wound tightly and without spaces between the turns, as shown in Fig 2. It is tapped at each turn up to 10, then at 15, 20, 25, 30, 35, 40 and 45 turns. The use of crocodile clips permits any number of turns to be selected.

The taps on the coil are formed by bending the wire back on itself and twisting a loop. The loops are then scraped and tinned with solder. Some enamelled copper wire is self-fluxing and only requires the application of heat from a hot soldering iron bit loaded with fresh solder for a few seconds before it will tin. Make a good job of tinning the loops to ensure the crocodile clips make good contact. The capacitor is a polyvaricon from an old radio. I used a 2 x 200pF component with both gangs in parallel; other values will work.

The ATU was made up on a wooden base with scrap PCB for the front and rear panels, as in Fig 3. You will need a socket to go to the transmitter or receiver and one each for the wire aerial and the counterpoise or earth. You can use whatever matches your existing equipment. The prototype used a phono socket for the transmitter and two 4mm sockets for the aerial and

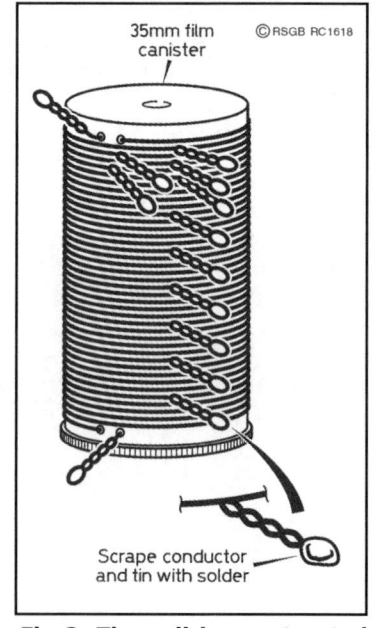

Fig 2. The coil is constructed on a 35mm film container.

AN L-MATCH ATU

earth. The coil is mounted by screwing the lid of the film container down on the base, then snapping the completed coil assembly into it. I glued the capacitor to the front panel. Wire up point-to-point, as shown in Fig 3.

OPERATION

When completed, you can check operation with a receiver or a low-power transmitter (less than 5W). With a transmitter, use a VSWR meter to find the coil tap which gives the lowest VSWR, then adjust the capacitor to tune to minimum VSWR. Incidentally,

Front view of the simple L-match ATU, showing crocodile-clip connection to the coil. The knob must be of the insulated type, as both sides of the variable capacitor are potentially live

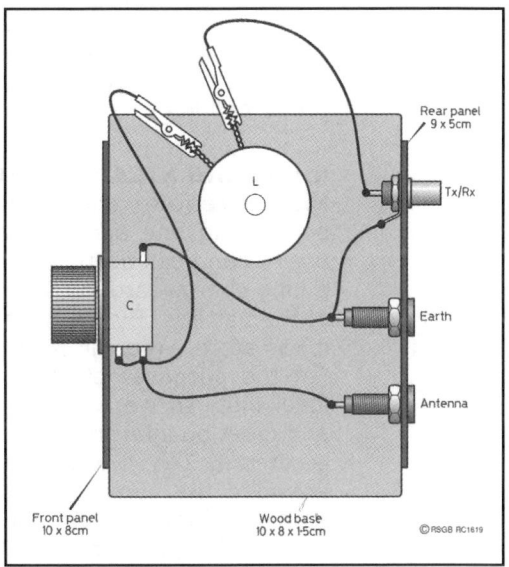

Rear view of the ATU, showing sockets and connections to the capacitor.

remember not to touch any exposed metal within the ATU while on transmit! An insulated knob and a scale help you to note the position of the capacitor for future reference. To set it up using a receiver only, find the best position for the tap and the capacitor by listening to a weak signal and adjusting for the loudest signal (with the help of an S-meter, if you have one).

Fig 3. Physical layout. The front and rear panels are fixed to the base with wood screws.

As a guide, I used 20 turns on 160m, 10 turns on 40m, 4 turns on 20m, and 2 turns on 10m and 15m.

A CHARGER FOR NICAD BATTERIES

The cost of replacing dry batteries can be alleviated to a great extent by use of rechargeable batteries such as those containing nickel and cadmium (called NiCads). They are more expensive than the batteries they replace, but they can be charged hundreds of times and thus prove cheaper in the long run.

NiCads produce only 1.2V per cell, compared with the 1.5V of standard cells, so before you rush out and buy lots of these, please make sure that the equipment on which you plan to use them will work on the lower voltage! For example, if you are using four cells to give you a 6V supply, NiCads will give you only 4.8 V, which is quite a reduction. Using six cells to replace a 9V battery will give you only 7.2V. Not all equipment is happy with these reductions!

However, we all use them when we can, and they save substantial amounts of money. Here is the circuit of a charger to keep them in prime condition.

CHARGING NICADS – THE AMPERE-HOUR
NiCads require charging at constant current, which means that connecting one across a normal power supply (constant voltage) is useless and can destroy it. It needs pampering to the extent of needing a long charge (around 16 hours) at a rate dependent upon the capacity of the battery. By the capacity of a battery, we mean how much energy it can store. You will probably know that energy is measured in joules. For the purposes of storing energy in batteries, the joule is not the ideal unit, so we use one that is! This unit is the ampere-hour (Ah), and must be interpreted with some realism. For example, if the battery is rated at 2Ah, it will deliver a current of 0.5A for 4h, or 0.25A for 8h. Provided the current is not too high, the product of the current (in amps) and the time (in hours) for which it will flow before the battery is flat will always be around 2Ah. A workable 'rule of thumb' for calculating the charging current is that its value should be around one-tenth of the numerical value of the capacity; so, for our 2Ah battery, a charging current of around 200 mA (2 ÷ 10 = 0.2A or 200mA) would be used.

CONSTANT VOLTAGE TO CONSTANT CURRENT
Many integrated circuit (IC) chips are available for use as voltage regulators, ie they supply a constant voltage. Most of these can be persuaded to become constant current supplies with one external resistor!

A CHARGER FOR NICAD BATTERIES

The voltage regulator IC usually has only three connections – 'input', 'output' and 'common'. It is designed (in the case of the LM780S) to produce a constant 5V output between the 'output' and the 'common' connections, at currents of up to 1A. If a resistor is connected between these, the IC will maintain 5 V across it. If you look at Fig 1, you will see the circuit performing this conversion.

For the previous example, we derived a charging current of 200mA, so we now need to calculate the value of resistor that will produce this current. Using the equation which is derived from Ohm's law:

$$I = \frac{V}{R}, \text{ from which } R = \frac{V}{I},$$

where R = R1, the resistance in ohms that we are calculating,
       V is the voltage across R1 (5V), and
       I is the current flowing (200mA).

So,

$$R1 = \frac{5}{0.2} = 25\Omega$$

25Ω is not a 'common' or 'preferred' resistor value, so we must choose the next largest value, which is 27Ω. This reduces the current, but only slightly – it is now 185mA. When calculating resistor values in power supply circuits, we must always check the powers that they dissipate and make sure we specify and fit suitable resistors.

Power (in watts) is the product of the voltage across and the current through a device, so in this case it is given by:

Power = V x I = 5 x 0.185 = 0.925W.

Rather than use a 1W resistor operating very near its limit, it is safer to use a 2W resistor operating well within its limits.

Looking again at Fig 1, we now have a constant-current source producing 185mA, when R1 is a 27Ω, 2W resistor. For use with NiCads requiring charging currents other than 200mA, you will need to repeat the two equations above, using a new value for I.

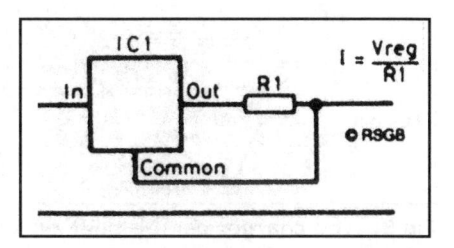

Fig 1. Voltage regulator arranged to produce a constant current.

THE FULL CIRCUIT AND ITS ASSEMBLY

This is shown in Fig 2, and can be broken down into three parts. The first is the voltage conversion produced by the mains transformer, T1. The second part, the bridge rectifier, BR1, and the smoothing capacitor, C1, convert the AC to smoothed DC. The third is the constant-current section already discussed.

WEEKEND PROJECTS

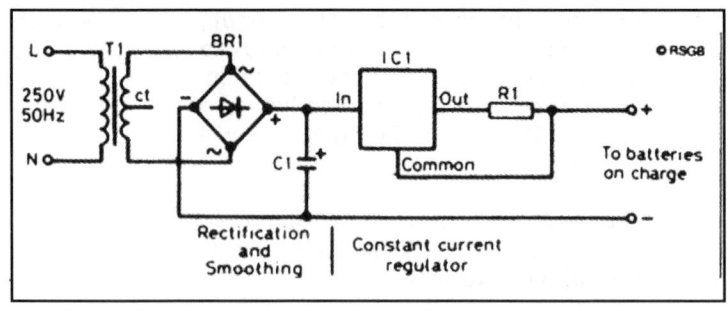

Fig 2. The complete circuit, including mains transformer.

The prototype was assembled on matrix board measuring 18 holes by 12 strips, although, as Fig 3 shows, this is much larger than is strictly necessary. No strip cutting is needed, but make sure that IC1 and C1 are inserted correctly.

Warning! Before you attempt to wire up the transformer and the bridge rectifier, be aware that you will eventually be connecting the circuit to the mains supply, so there are three possibilities for you: (1) get a qualified friend to supervise your completion and testing of the circuit; (2) get your qualified friend to complete and test the circuit for you; (3) replace the transformer and bridge rectifier with a mains adapter.

If you decide to use the mains adapter, its output should be connected directly across C1, because T1 and BR1 are now no longer needed. Make sure the polarity (positive and negative) is correct and that the adapter output is set to 12V.

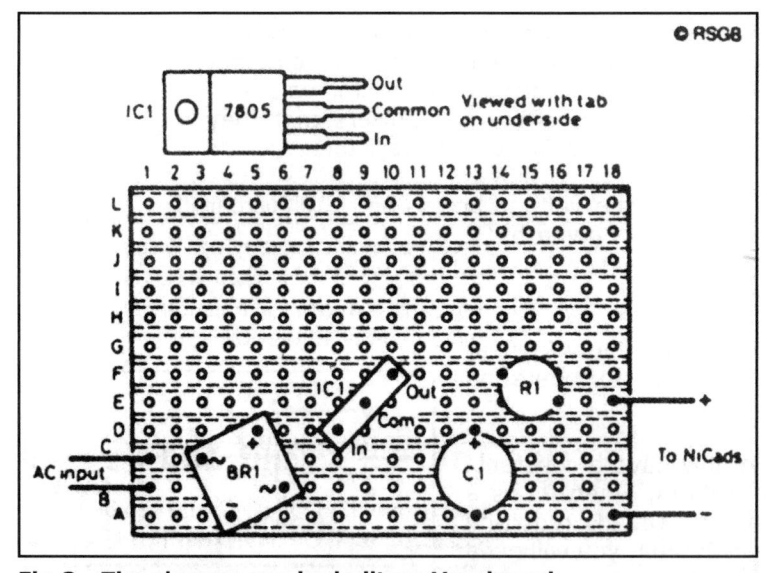

Fig 3. The charger can be built on Veroboard.

A QUICK TEST

If you have built the mains version, make sure all connections are correct, and that there are no soldered joints which will touch other parts of the circuit. The box must be securely closed before tests begin. The RSGB cannot be held responsible for damage to equipment or batteries! The version using the mains adapter need not be closed during tests.

With nothing connected to the output, switch on. The unit should run cold. If this is not the case, switch off quickly and recheck your circuit. If all is well, switch on again and connect a DC multimeter (on the 'current' range) across the output. It should indicate only a slight difference from the calculated value of 185mA. You can now charge your NiCads!

A CHARGER FOR NICAD BATTERIES

Other charging currents can be set by having different values for R1, perhaps selectable by a rotary switch. Remember to make sure that the values of both resistance and power dissipation are correct, and don't exceed the 250mA rating of the transformer (or the 1A rating of the IC if you are using a bigger transformer).

PARTS LIST

		Maplin code
Resistor		
R1	As required – see text	
Capacitor		
C1	1000µF, 35V electrolytic	FF18U
Semiconductors		
IC1	LM7805, 5V, 1A regulator	CR14Q
BR1	W005, 50V, 1A full-wave rectifier	QL37S
Transformer		
T1	9-0-9V, 250 mA, sub-min transformer	YN15R
Additional items		
Case		FG41U
Veroboard, 18 holes by 12 strips		
Plug to suit NiCads		
Double-sided sticky tape as required		
Insulated wire for battery connections		

AN OP-AMP TESTER

When building a circuit, it is not unusual to find that it doesn't work first time. After fault finding it's not unusual either to find that one of the supply rails had been connected to either an input or output of an operational amplifier (op-amp). This could be either because of an incorrect link, an unnoticed short-circuit between copper tracks, or a direct connection between the IC pins due to a 'whisker' of solder. With a stripboard project it could also be due to an intended break in one strip that has not been completely cut through, or a burr of copper that is shorting to an adjacent track. It might also be due to a required break that has been omitted.

When the fault has been rectified and the circuit still doesn't work, it then isn't possible to say whether this is now due to another fault present or whether it is because the op-amp was damaged by the initial fault. The best solution to this dilemma is to check independently that the op-amp is working correctly. This simple project will perform just such a check. It can also be used to check that an op-amp salvaged from unwanted equipment for use in another project works correctly. It will prevent time being wasted in checking for construction faults, when it is in fact the component that is faulty.

HOW IT WORKS
When a component fails, particularly a semiconductor device, it usually fails catastrophically, ie it fails completely and doesn't just work half-heartedly. In the case of an op-amp, this usually means that the output goes to the value of one supply rail or other and stays there. Another possibility is that the output goes to some fixed DC value between the two supply rail limits.

This circuit works by incorporating the op-amp under test into an astable oscillator circuit. If the circuit oscillates, the op-amp is fine; if not, it is damaged. In order that an oscilloscope is not needed to observe the output waveform, the op-amp output is connected to a detector circuit and the output from this is connected to an indicator section. The indicator section drives two LEDs: a green one to indicate that oscillation is present, ie the op-amp passes the test; and a red one to indicate no oscillation, ie the op-amp fails the test.

THE CIRCUIT
Components R1 to R3 and C1, in combination with the op-amp under test, form an astable oscillator (see Fig 1). When this is operating correctly, the output Vtest is a square wave with a frequency of about 1kHz. Otherwise Vtest will be a DC voltage.

AN OP-AMP TESTER

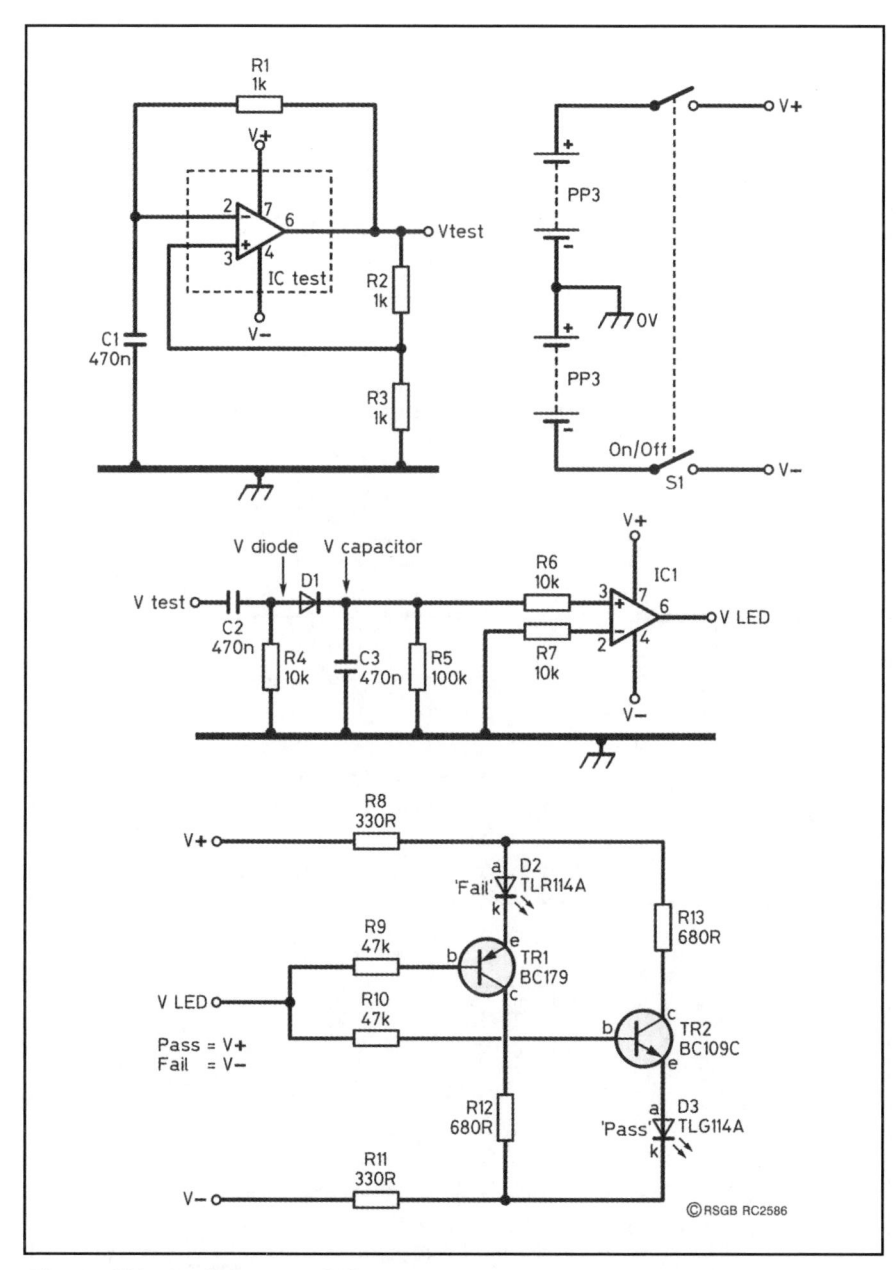

Fig 1. Circuit diagram of the op-amp tester.

If *V*test is a DC voltage it will be blocked by C2 and *V*diode will fall to 0V as the right-hand plate of C2 discharges via R4 (with time constant C2 × R4). D1 then isolates IC1 from this part of the circuit and both inputs of IC1 are connected via resistors to 0V. The non-inverting ('+') input, however, is connected via 110kΩ (R5 + R6), whereas the inverting ('−') input is connected via 10kΩ (R7). This means that the bias current entering the '−' input will be greater than that entering the '+' input,

WEEKEND PROJECTS

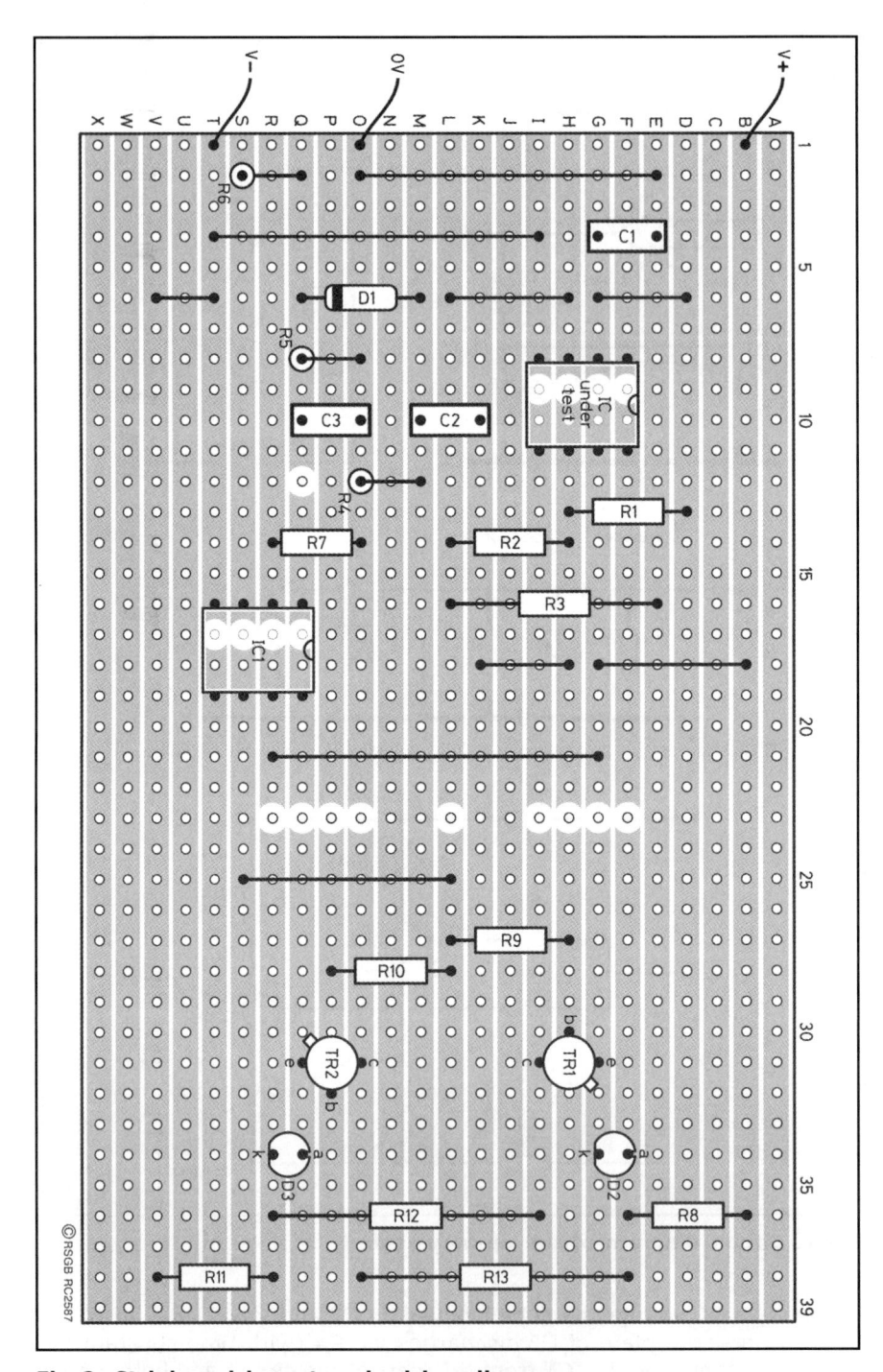

Fig 2. Stripboard layout and wiring diagram.

deliberately creating a differential input offset voltage. Because of this, and the high gain of an op-amp connected without any negative feedback resistors, the output of IC1, VLED, will very nearly swing to

AN OP-AMP TESTER

the negative supply voltage. This switches TR1 on and TR2 off, and so the 'Fail' LED D2 lights.

If, however, Vtest is a 1kHz square wave, C2 will not block the signal and D1 will conduct during the positive part of the waveform. This means C3 will charge up, as R5 is now acting as the discharge path for C3 and the time constant C3 × R5 is long compared with the period of the 1kHz signal. The resulting positive value of Vcapacitor causes the current flowing into the '+' input of IC1 to exceed the current flowing into the '-' input and VLED swings nearly to the positive supply voltage. This switches TR1 off and TR2 on, and so the 'Pass' LED D1 lights.

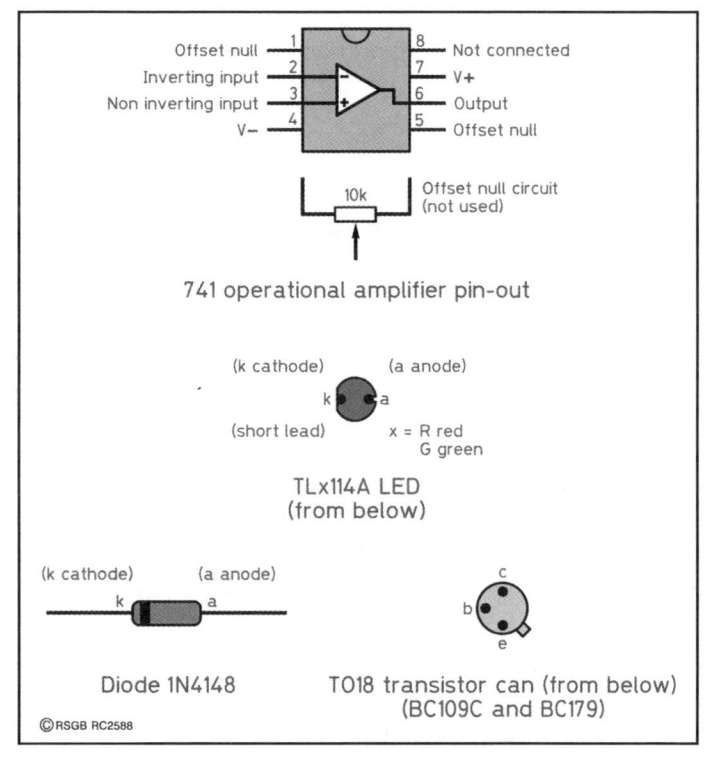

Fig 3: Orientation of critical components.

CONSTRUCTION

The stripboard layout for the op-amp tester is shown in Fig 2. An 8-pin DIL socket is of course needed for the op-amp under test, but IC1 can be inserted into a socket or soldered directly into place as preferred. It is always a good idea to install ICs in sockets though; as well as eliminating the possibility of damage due to soldering it makes removal for testing much simpler. Care needs to be taken to ensure that the op-amp, transistors, diode and LEDs are connected the right way round, and Fig 3 shows how to determine the correct orientations.

COMPONENTS

All the semiconductors used in this project are general-purpose types, and the values of the components associated with them are chosen to limit voltages and currents to their working values. Otherwise, values of capacitance and resistance are chosen to give suitable time constants to ensure reliable operation.

IN USE

Any single op-amp package with standard DIL pin-out connections that operates with a power supply of ±9V can be tested, which covers most situations. Simply

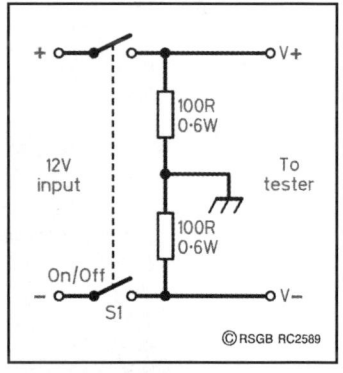

Fig 4: To power the Op-Amp tester from a 12V supply, substitute the top right hand part of Fig 1 with this circuit. If you do this, remember that the 'chassis' of the tester (denoted by the chassis symbols) will not be at ground potential, so you will need to keep the unit isolated from ground.

121

WEEKEND PROJECTS

insert the op-amp to be tested into the test socket, again ensuring correct orientation, switch on, and note which LED lights.

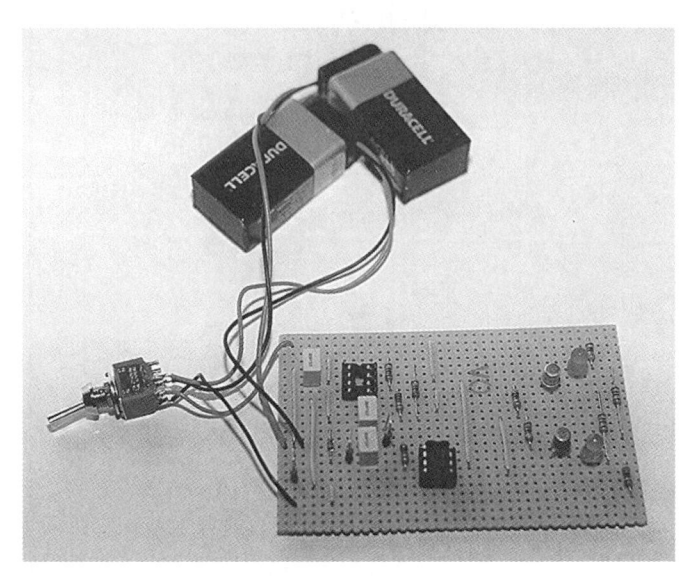

USING A 12V SUPPLY

If the use of two 9V batteries to power the op-amp tester doesn't appeal to you, it is possible to adapt it to run from a 12V power supply. To do this, simply take two 100Ω resistors, make a potential divider (as shown in Fig 4), and use this instead of the top right-hand part of Fig 1.

The completed op-amp tester.

PARTS LIST
Resistors - All resistors are metal film, 0.6W 1%
R1 – 3	1k
R4, 6, 7	10k
R5	100k
R8, 11	330R
R9, 10	47k
R12, 13	680Ω

Capacitors
C1 – 3	470n polyester film

Semiconductors
IC1	LM741CN
TR1	BC179
TR2	BC109C
D1	1N4148
D2	TLR114A red LED
D3	TLG114A green LED

Additional items
S1   DPDT
Stripboard
PP3 battery and clip – both × 2

OPTICAL COMMUNICATION

Many principles of radio communications can be demonstrated very effectively by a wireless link using visible light. This is not surprising, since both light and radio waves are types of electromagnetic radiation, but light allows us to see what is actually happening. The system described here is a simple optical transceiver which has been a successful project for students and teachers in radio clubs at a number of local schools.

DESCRIPTION

The project uses an LED and photodiode, and is based around an LM324 'quad' operational amplifier (ie four op-amps in one package). The circuit diagram is shown in Fig 1 and the PCB foil pattern in Fig 2. The transmitter uses one of the op-amps, the receiver uses two, and there is one spare. Op-amps amplify any voltage difference between an inverting and non-inverting input, marked '-' and '+' respectively. In this system, the non-inverting input ('+') is used as a reference point at half the supply voltage, so any signal on the other input is amplified in comparison to this reference voltage. Amplifier gain is determined by the ratio of value of the feedback resistance to the input resistance (eg R1/RV1).

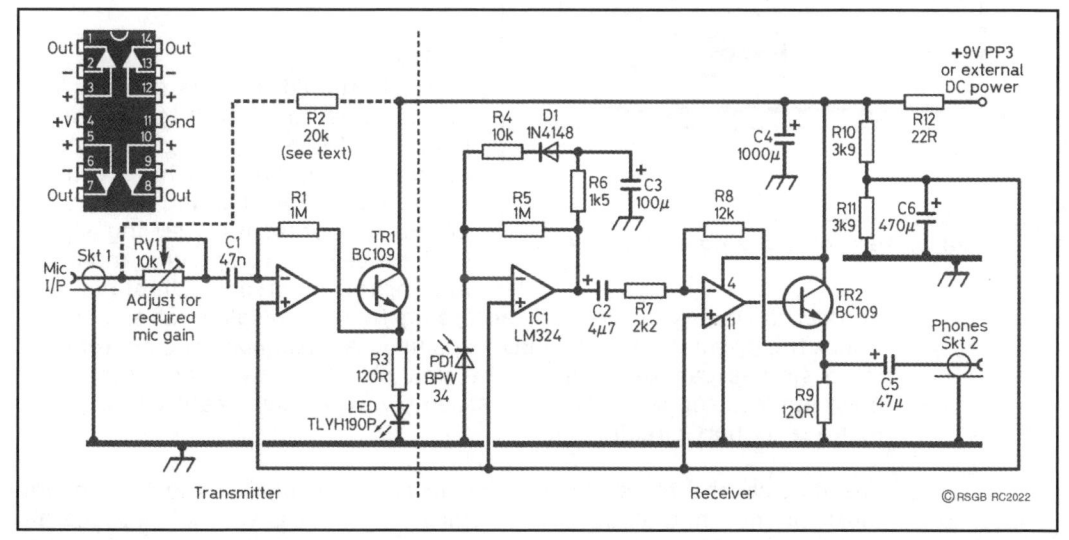

Fig 1. The low component count of the optical transceiver is largely due to the use of the LM324, a quad op-amp.

WEEKEND PROJECTS

The preferred LED produces a very bright light with a narrow beam. Other LEDs may be significantly less powerful and will have more limited ranges. It is possible to use a photocell from a model kit as the detector, but the photodiode and LED can be obtained from a good component supplier (for example, Farnell Components, **www.farnell.com**). R10, R11, R12, C4 and C6 are included to ensure supply voltage stability.

In the transmitter, the LED glows continuously, since the transistor is turned half on by the op-amp. When a small signal is produced by the microphone, it is amplified by the op-amp which tries to turn the transistor on and off, making the LED flicker in sympathy with the input signal. Hence we have amplitude-modulated visible light. If you use a dynamic microphone, the pull-up resistor R2 *must be* omitted, but if you use an electret microphone it *must be* included.

The receiver begins with a photodiode, which varies in resistance according to how much light is present. Changes in resistance are amplified by an op-amp, but since this is still quite a small signal, a second op-amp is used to amplify the signal further. Again, a transistor is turned half on by the final op-amp, driving headphones as the audio output.

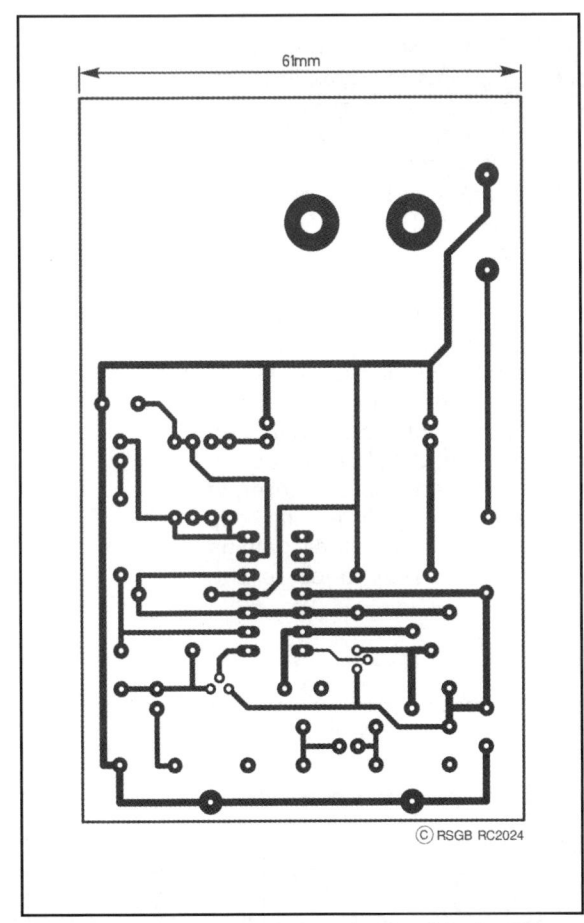

Fig 2. Foil pattern of the single-sided PCB.

TESTING

The LED should glow as soon as the unit is turned on and, if testing takes place in the presence of domestic mains lighting, a loud hum will be heard because AC lighting 'flickers' at 100Hz and presents a strong signal to the receiver. TV screens, monitors and other electronic displays will produce a variety of sounds, according to their display refresh rates. Check the orientation of the LED if it is unlit, and similarly the photodiode if no signals can be heard. The unit should draw about 45mA, so anything less suggests incorrect assembly, while significantly more may indicate a short-circuit.

The unit can be tested in isolation using a mirror, but the light beam must hit the photodiode directly since a near miss will not produce an audible signal.

OPTICAL COMMUNICATION

EVALUATION AND EXPERIMENTATION
The optical transceiver can be used to demonstrate many aspects of radio theory.

Amplitude modulation (AM)
Although the LED is modulated by an audio-frequency signal from the microphone, the light is only ever seen to flicker if distortion occurs by overmodulation.

A more effective demonstration of AM is achieved by interrupting a beam of light with a fan, rather than modulating it at source. Any light source will do and although an electric fan can be used, a cardboard propeller on the end of a hand drill illustrates the principle most effectively.

Automatic gain control (AGC)
Radio receivers often include an automatic gain control, which reduces their sensitivity when strong signals are present. This receiver circuit includes a very simple automatic gain control, based on D1. If the voltage across the feedback resistor is less than 0.6V, this is insufficient to forward bias the diode, and feedback through the 1MΩ resistor R5 assures high gain. If the voltage across the feedback resistor exceeds 0.6V, the diode becomes forward-biased and the resistance of the feedback loop becomes less than 10kΩ, reducing the gain significantly. The circuit will work without this feature, but is easily silenced by strong background light, which can be demonstrated by temporarily removing D1.

Selectivity
Selectivity is the ability of a receiver to select one signal in preference to others. Although the photodiode is most sensitive to orange light, it will also detect light of other frequencies, which are visible as other colours. In fact, it will even detect infra-red signals invisible to the naked eye, so the unit can 'hear' the commands of a TV remote controller. The output of the Toshiba LED has significant infra-red content, so infra-red filters can be included to create an invisible link. Simple filters using coloured cellophane can be used to experiment with a selection of signals from different coloured lights. Offcuts from lighting filters used in local theatres may be obtainable from the stage lighting manager.

Noise filtering
It is evident from the background hum of the receiver in the presence of AC lighting that AM signals are susceptible to interference from other signals. In this case, a simple 200Hz high-pass filter in series with the headphone output can be used to attenuate the low-frequency mains hum without blocking the wanted audio signal. This can be achieved using a simple RC filter, as shown in Fig 3. The nominal cut-off frequency is given by the formula below, and the effect on the signals heard is quite noticeable.

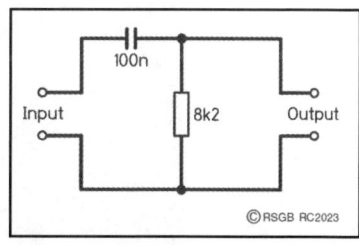

Fig 3. Simple high-pass audio filter to reduce mains hum.

WEEKEND PROJECTS

$$f = \frac{1}{2\pi RC},$$

where f (Hz) is the cutoff frequency,
         R (ohm) is the value of the resistor in Fig 3, and
         C (Farad) is the value of the capacitor in Fig 3.

DX CONTACTS
Power and aerials are the major considerations for DX enthusiasts. The LED specified is actually rated at 50mA, so the value of R3 can safely be reduced to 39Ω to increase power output (R12 then needs to be $1/_4$W). Both receiver and transmitter performance can be significantly improved by the use of lenses or a magnifying glass. This focuses the signal carrier, just as directional aerials do in radio communications. A more focused beam can be obtained if the cap of the LED is cut off (eg with a junior hacksaw) and the LED then polished on an abrasive stone and finished-off with Brasso on newspaper. The LED can then be mounted at the focal point of a lens. Do not shine the LED directly into anyone's eyes.

Experiments have achieved distances approaching 100m.

FURTHER INVESTIGATIONS
If the photodiode is connected to a crystal earphone with no other components, it will detect strong signals such as neon or fluorescent lighting in close proximity. Like a simple crystal radio receiver, no batteries are required, but lenses will improve performance significantly, just as a good aerial will for a radio receiver. Other ideas could be to build a repeater, by taking the receiver output directly to the transmitter input, or to use a comparator interface to allow data and SSTV to be sent from a computer. The system can be used as a dedicated audio link or as an educational aid, but in either case it provides an excellent opportunity to understand some of the principles of telecommunications.

PARTS LIST
Resistors - All resistors 0.1W 5% tolerance, unless otherwise stated

R1, 5	1M
R2	20k
R3, 9	120R
R4	10k
R6	1.5k
R7	2.2k
R8	12k
R10, 11	3.9k
R12	22
RV1	10k preset potentiometer

Capacitors

C1	47n
C2	4.7μ, 16V

OPTICAL COMMUNICATION

C3	100μ, 16V
C4	1000μ, 16V
C5	47μ, 16V
C6	470μ, 16V

Semiconductors

IC1	LM324
TR1, 2	BC109
D1	1N4148
LED	TLYH190P (Toshiba)
PD1	BPW34 photodiode

Additional items
Battery clip
Jack socket(s)

PACKET RADIO PRINCIPLES

Following the end of the Second World War, there was a glut of surplus teleprinters on the market. Many of these machines (and their successors) were made by the venerable Creed company in Croydon. It wasn't long before these noisy, smelly and utterly lovable machines were being used for RTTY (radio teletype) operation on the amateur bands. Using an old Creed machine was preceded by a fair amount of construction of terminal units and modification or repair of the machine. This was the beginning of the world of data communications for the amateur. Prices were low and the satisfaction of working this mode was immense. As the affordable home computer came on the market, the old Creed machines were gradually replaced. Little did we know that an even bigger change was soon to take place, as the packet network was born and slowly became used worldwide. RTTY is still going strong of course – the modern computer and its versatile sound card have seen to that!

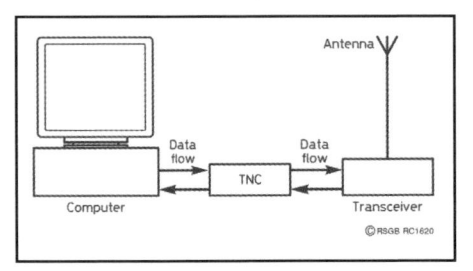

Fig 1. The constituent parts of a typical simple packet station.

THE ORIGINS OF PACKET

Packet radio stems from a method (or *protocol*) used to set up communications links between computer networks. The protocol is known as *X25*. The amateur version had some changes made to it and is known as *AX25*, the 'A' standing for 'amateur'.

Packet radio is a method of data communication; it allows one computer to talk to another via radio. Obviously, there must be a way of connecting the digital signals in the computer (strings of 1s and 0s) to the transceiver. This is accomplished by a *terminal node controller* (TNC). This handles all communications between the computer and the transmitter and (in the reverse direction), between the receiver and the computer. Fig 1 illustrates this. In effect, the TNC is a type of modem. It takes the digital signals from the computer serial port (RS-232C) and turns them into sound signals that can be transmitted by the rig (Fig 2).

The TNC also handles the way in which the link works, sending control signals and instructions to the TNC at the other end. The only other thing to remember is that you will need some software to allow you to communicate with the TNC. There are many packages available, many of them shareware and freeware, so do look around until

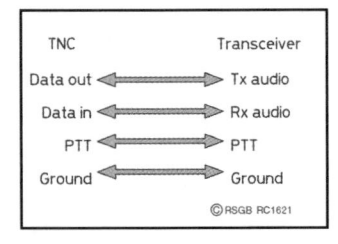

Fig 2. The TNC-to-transceiver link.

PACKET RADIO PRINCIPLES

you find something you like. You may like to know that any 'terminal' program will allow you to send simple alphanumeric data to the TNC, but the specialist programs have many useful features included.

HOW IT WORKS

Before starting data transfer, the TNC must 'connect' with another station. This is done by issuing the connect command to the TNC, followed by the callsign of the station at the other end of the link. For example, to connect to G9ZZZ one would type 'Connect G9ZZZ'. This is normally abbreviated to 'C G9ZZZ'. Once the connection is made (the message 'CONNECTED TO G9ZZZ' appears on the screen), data transfer can begin.

So how does packet work? The user types his message at the computer keyboard and the TNC assembles the data coming from the computer into little 'packets'. Each packet contains the callsign of the destination station, the callsign of the originating station, plus control and error detection information. This is in addition to the data actually being sent by the operator. The packet is then sent to the transceiver for transmission. If the data is received correctly, it is acknowledged and the next packet is sent. If the sent packet is corrupted, the receiving station will request a repeat. Theoretically the repeat requests could go on indefinitely but the number of repeats is set by the user, and normally 15 retries is the limit.

There are three data transmission rates in use at the moment. These are 300bps (bits per second or baud), which is normally used on HF; 1200 and 9600 baud, used normally on VHF. This last speed normally requires modification of the transceiver without which the audio filters would not allow the signal to pass. There are some transceivers on the market designed to work at this speed without modification.

PACKET QSOs

How do you start a packet QSO? Well, in much the same way as you start a phone contact. Either call CQ, or arrange a QSO with a friend. Once connected to the remote station (remember the 'C G9ZZZ' command?) you can type text in at your computer, just as if you were talking to him or her. The responses from the other station will be displayed on the monitor for you to read.

A point about operating practice is worth mentioning here. When you have finished sending your latest sentence(s) use chevron signs (>>) to denote the end of a transmission. It is also important to remember to wait for the chevrons from the other end before you start sending, otherwise things can get a little confusing.

Most terminal programs send the text to the TNC at the end of a line. You can send it at any time you like by just hitting CR (carriage return). A point about station identification when using packet: if your transmissions last longer than 15 minutes, you *must* send either a CW or phone ID. Please look in the *Amateur Radio Licence Terms, Provisions and Limitations Booklet BR68* for more information.

WEEKEND PROJECTS

BULLETIN BOARDS

As well as connecting to another station and having a chat, it is also possible to connect to a *bulletin board* (its abbreviation 'BBS' stands for 'bulletin board station' or 'bulletin board system'). A bulletin board can be connected to in the same way as a 'normal' station but most of the time there is no operator present. The SysOp (system operator) provides a station that is connected to the packet network. The BBS stores messages from other amateurs. These messages can be either personal mail to a given station or they can be general messages (*bulletins*) intended for a general audience. You will find all sorts of useful files on BBS computers. The files can be accessed and downloaded by your packet station. Mail for both you and for other stations can be sent to or from a BBS, and bulletins can be read (and you can get into some interesting debates at times).

There are two ways of checking for personal messages and seeing what bulletins have been posted. The first involves actually 'logging on' to the BBS. All BBSs will tell you when you log on if there are messages waiting for you. You can then read the mail when you want to. One thing here: please remember to 'kill' your mail once you finish with it. If you don't it will stay on the system, clogging up memory for no reason. You can imagine what would happen if we all just left our messages in memory. You can look at bulletins by scanning the BBS message list and downloading the messages you want directly. Alternatively, you can download the message list directly to your computer. Then you can decide which (if any) you want to read and then collect them later.

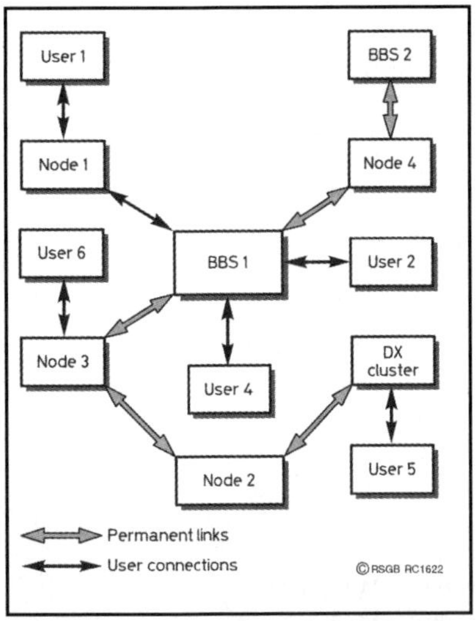

Fig 3. A typical packet network. A DX Cluster is a special sort of packet station set up to allow users to share up-to-the-minute DX information.

THE PACKET NETWORK

In one way, packet radio is probably one of the politest modes in the amateur service: when several stations are using the same frequency – as often happens – the various TNCs take it in turn to transmit. This reduces the amount of collisions (two stations transmitting at the same time). This will reduce the data transfer rate, as the number of transmitted packets is reduced, but this leads to greater spectrum efficiency.

One question you may be asking is "Where do all the messages come from?" Well, they actually come from other amateurs all around the world. It is not unusual to see messages from the USA or even further away. They travel here over *the packet network* (Fig 3).

The packet network is the 'Internet' of the amateur radio world. Each BBS is connected to the network, and hence to every other BBS. This allows them to access the fast links that run up and down the country so that messages can be passed throughout the UK. Some of these links are at microwave frequencies, others at UHF. Other countries have their own networks which work on the same principles as the UK one. All of these networks are linked together by HF and satellite 'gateways', allowing messages originating in the UK to travel to other parts of the world.

Packet radio may be found on most bands, but probably the most popular is 2m. Following the recent changes to the 144 – 146MHz band plan, digital modes have moved to the 144.800 – 144.990MHz sub-band. On HF, you will find digital modes just above the CW portions of the bands.

A word of warning: if you are using 300-baud packet on the HF bands, please avoid the beacon sub-bands, as interference can make beacon observation difficult.

YOUR FIRST PACKET STATION
Setting up a packet station need not cost the earth. I have seen TNCs that cost less than £70, free software, and an old IBM-XT or similar PC for under £20. You must remember that the PC is only being used as a terminal in this case. Later, you may want to move up to fax, AMTOR or SSTV, which will cost only a little more.

So how does it all go together? You will need to make (or buy) a cable to go between the PC and the TNC. Make sure that you get the right number of pins. Some computers have RS-232C outputs with nine pins, while others have 25 pins (you will have to look). Once you've linked the PC to the TNC, you must make up a lead between the microphone input and audio output sockets of the rig to the audio output and audio input sockets, respectively, on the TNC. Once this is done you are ready to use packet. Just load the software, fire everything up and you're ready to go – or almost.

Some TNCs have parameters that must be set up. Here, I am afraid that you will need to read the instruction manual to see what you need to do. You may well find that the manual gives you a lot of information, so it is well worth reading through. However, all TNCs have a 'default' set of parameters which should work first time. You can refine their values once you become knowledgeable about their effects.

Before you rush off, buy the kit and have a go, can I make a suggestion? There are several good books on the market about packet. Look at the books listed below and you will find all the help you need. The British Amateur Teledata Group (BARTG) is also a good information source, and a contact address is given.

Good luck, and enjoy your packet operations.

REFERENCES
[1]   *Your First Packet Station*, G0WSJ, RSGB.
[2]   *Packet Radio Primer*, G8UYZ and G8NZU, RSGB.
[3]   BARTG membership enquiries: Bill McGill, G0DXB, 14 Farquhar Road, Maltby, Rotherham S66 7PD. E-mail: members@bartg.demon.co.uk.

A PORTABLE POWER SUPPLY

When time permits, I enjoy participating in the Backpacker series of RSGB contests [1], operating in the 3W category. The power supply I use for these outings is described here. It is ideal for the purpose, and can also be used as a simple uninterruptible power supply (UPS) for the shack.

DESIGN CRITERIA
- A maximum of 4.5kg (10lb) in weight.
- Able to be carried in a small rucksack.
- Able to be plugged into the nearest mains outlet to be recharged.
- When at home to run in 'float-charge mode', to operate low-current equipment.
- To be of reasonable cost.
- To use readily available components.

At the 1998 Rainham Rally I found a couple of sealed lead-acid cells rated at a nominal 12V and 7Ah capacity. They weighed in at a little over 2.2kg (5lb) – just what I needed!

Being of the sealed variety, care has to be exercised in not overcharging the cell, ie to prevent gassing, so 13.8V would be the maximum permitted voltage at the terminals. Thus a stabilised supply was essential.

THE CIRCUIT
The float charger is hardly original – indeed it was adapted from that excellent series of articles by John Case [2]. One major consideration was the need to 'over-engineer' it, as I would leave it plugged in and switched on almost continuously.

Fig 1 shows the circuit diagram. Many of the components were salvaged from redundant equipment or the junk box; however, the critical components – transformer, pass transistor, reservoir capacitor and regulator chip – were all purchased new.

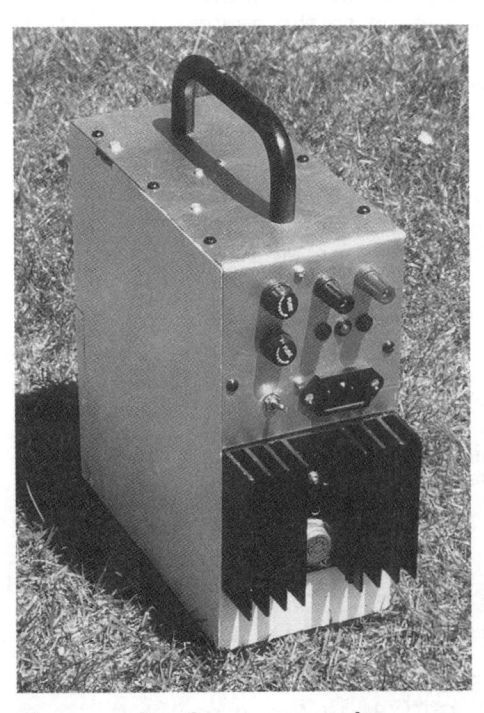

The completed power supply.

A PORTABLE POWER SUPPLY

There are many options available for layout; mine were dictated by the size and shape of the heatsink. In my case, the box that holds it all is made from 12mm plywood, with the major components mounted on an L-shaped aluminium plate which forms the front and part of one side.

A voltage-dependent resistor (VDR) is fitted across the primary of the transformer, and an over-current control is provided to limit the current to 1.5A.

A heavy-duty diode D4 is incorporated in the feed to the battery, and LEDs are fitted at strategic points as a confidence feature and for ease of fault-finding. The voltage controller, an LM723, is fitted, along with its components, on a small piece of matrix board, the remaining components being wired 'point-to-point' using substantial cable for the heavy-current paths.

Setting-up is quite straightforward – connect everything together to a fully-charged battery, connect a voltmeter to the battery's terminals, turn on and adjust the output to 13.8V as measured at the battery terminals. Set the current
limit to 1.5A, which in practice appears rarely achieved.

RESULTS
Does it meet the criteria I set out earlier? That I will leave you to judge. The figures are:

- Weight – 11lb.
- Size – 25 × 25 × 10cm, including heatsink and handle.
- Cost – around £20.
- Capacity – enough for a full Backpacker session

Even when left on continuously, the temperature of the portable power pack hardly rises above room temperature.

REFERENCES
[1] 'Backpacking – summertime delights', G6TTL, *RadCom* May 1997.
[2] 'Power supplies on a shoestring', GW4HWR, *RadCom* July – August 1986.

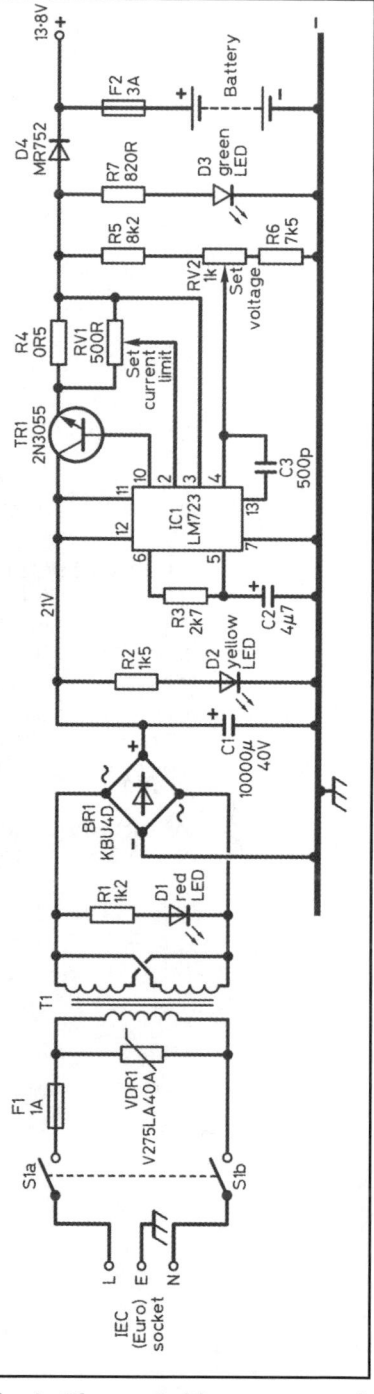

Fig 1. The portable power-pack works by float-charging a sealed lead-acid battery, to provide an uninterruptible supply.

133

PARTS LIST

Resistors - All resistors ½W metal film, 5% tolerance, unless otherwise stated

R1	1.2k
R2	1.5k
R3	2.7k
R4	0.5, 2W
R5	8.2k
R6	7.5k
R7	820
RV1	500 linear preset potentiometer
RV2	1k linear preset potentiometer
VDR1	V275LA40A

Capacitors

C1	10,000µ electrolytic, Maplin LE03
C2	4.7µ electrolytic
C3	500p ceramic

Semiconductors

BR1	KBU4D (or similar)
D1	Red LED
D2	Yellow LED
D3	Green LED
D4	MR752 (or similar)
IC1	LM723
TR1	2N3055

Additional items

F1	1A plus holder
F2	3A plus holder
S1	Double pole, single throw (DPST) toggle
T1	Mains transformer with two 15V @ 0.75A secondaries (Maplin DH27)

7Ah sealed lead-acid battery
IEC socket
Matrix board
Screw terminals, insulated
Case to suit

AN AMPLIFIED RF PROBE

When constructing radio-related projects, an RF probe is an item of equipment which is very useful to have around. It is also something simple to build. This RF probe employs a field-effect transistor (FET) amplifier to increase its sensitivity.

BACKGROUND

RF probes are often built to be used in conjunction with a multimeter. This ties up the multimeter, which cannot then be used to make other measurements at the same time. This simple project, the circuit of which is shown in Fig 1, adds an FET amplifier to the basic probe. The whole project is housed in a small metal box, complete with 9V battery.

The meter required is not critical in nature. The prototype used a scrap item from a tape deck, and had a full-scale deflection (FSD) of 200µA.

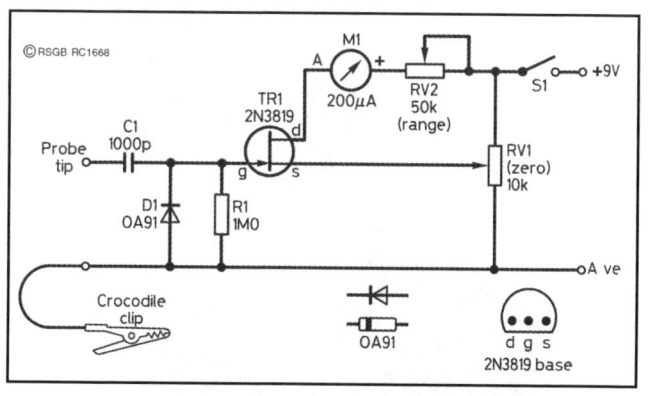

Fig 1. The amplified RF probe uses D1 to rectify an RF signal, TR1 to amplify it, then M1 to display it.

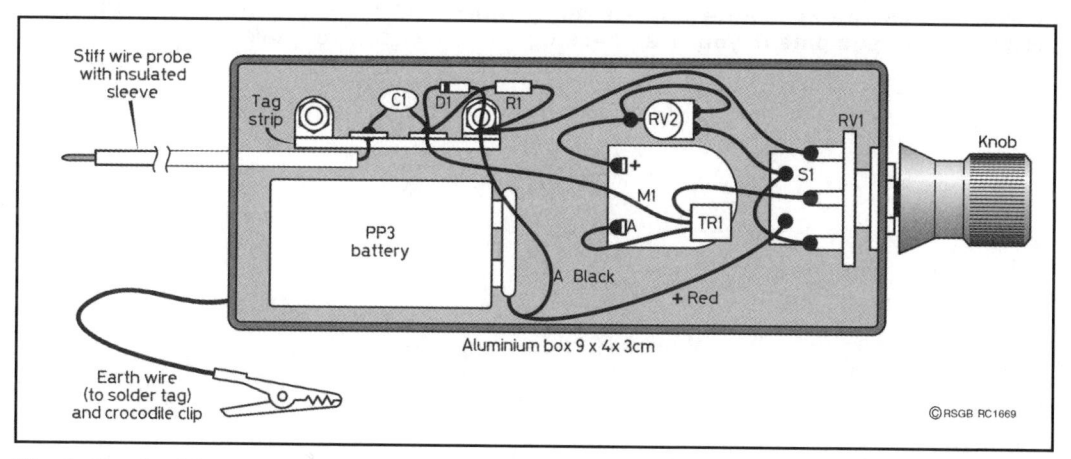

Fig 2. Physical layout. Note the insulation on the probe where it passes through the metal box.

135

WEEKEND PROJECTS

CONSTRUCTION
Components are wired up point-to-point, as shown in Fig 2. The probe is soldered to the tag strip, the insulation on it keeping it clear of the metal case. The on-off switch, S1, is part of a switched potentiometer (RV1), but you could just as easily use a toggle or other type of switch.

The probe itself is made from stiff copper wire. This needs to be covered with sleeving along most of its length, especially where it passes through the hole in the box. Diode D1 can be any small, germanium, point-contact diode, but a Schottky type would make the unit even more sensitive.

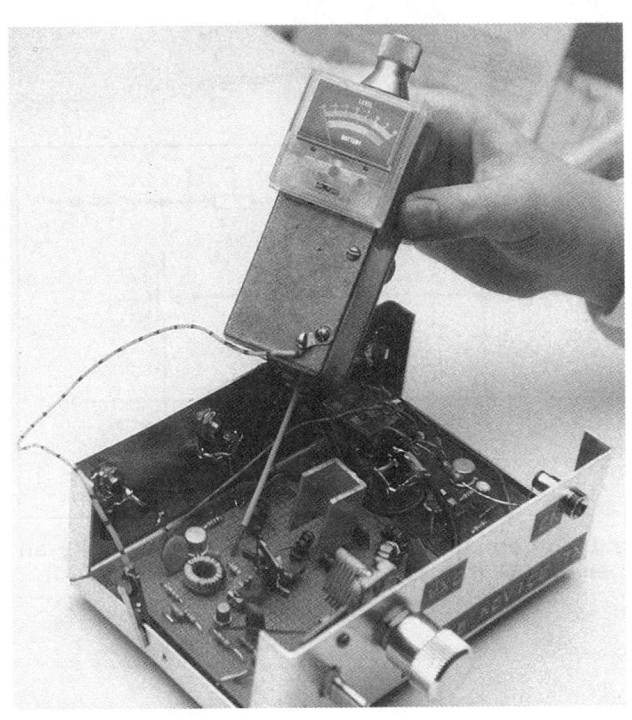

The photograph shows the connection of the short earth lead to a solder tag on the metal case. The crocodile clip on the end of this lead *must* be connected to the earth of the equipment being tested.

In use, attach the crocodile clip to the ground (earth) of the equipment you are checking.

OPERATION
Switch the probe on, then adjust RV1 so that, with no signal present on the input, the needle of M1 is just above the zero mark. With the probe touching an RF signal source, you should see the meter needle rise. RV2 is used to control the sensitivity of it. Although it would be possible to calibrate the RF probe against a known millivoltmeter, in most instances it is only required to either see whether a signal is present or not, or to adjust a circuit to peak a signal.

A SIGNAL INJECTOR

One of the most satisfying things in amateur radio is being able to locate and repair faults on equipment. One of the most useful tools for fault detection is a signal generator but, as these are often quite bulky and expensive, here is a simple, low-cost alternative.

DESCRIPTION

The circuit shown in Fig 1 is based around a two-transistor multivibrator, designed to produce square-wave oscillations at about 1kHz. Harmonics (at 3kHz, 5kHz, etc) ensure that the signal can be used not just to test audio circuits but RF circuits as well. When listening to the signal, it can best be described as a rather unpleasant buzz.

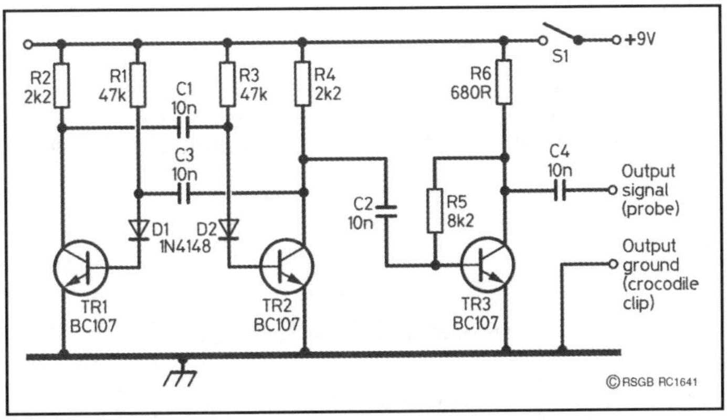

Fig 1. The simple signal injector is based on a multivibrator. This generates square waves, which are rich in odd harmonics.

The multivibrator consists of a two-transistor circuit in which the transistors are alternately turned on and off. D1 and D2 are used to ensure the multivibrator produces square waves with very 'sharp' edges, and hence the greatest harmonic content. In theory, the harmonics from this circuit continue to infinity but, in practice, there is a limit at which they can be detected. With the circuit shown the harmonics are detectable beyond 145MHz.

Output from the oscillator is coupled to an amplifier (TR3) via C2. The amplifier is used to ensure that any loading imposed by the circuit under test will not cause the multivibrator to stop. Biasing resistor R5 is low in value compared to many audio amplifiers but this ensures that the harmonic content of the output is as high as possible. Output coupling is via C4, which must be rated at 50V minimum to provide isolation and protection from the circuit under test.

CONSTRUCTION

The circuit is built on Veroboard (Fig 2), 9 strips × 30 holes. There are several track cuts, which can be made with a special track-cutting tool or a drill bit held in your hand. You can assemble the circuit in any order but, generally, the diodes and transistors are left to last. The

WEEKEND PROJECTS

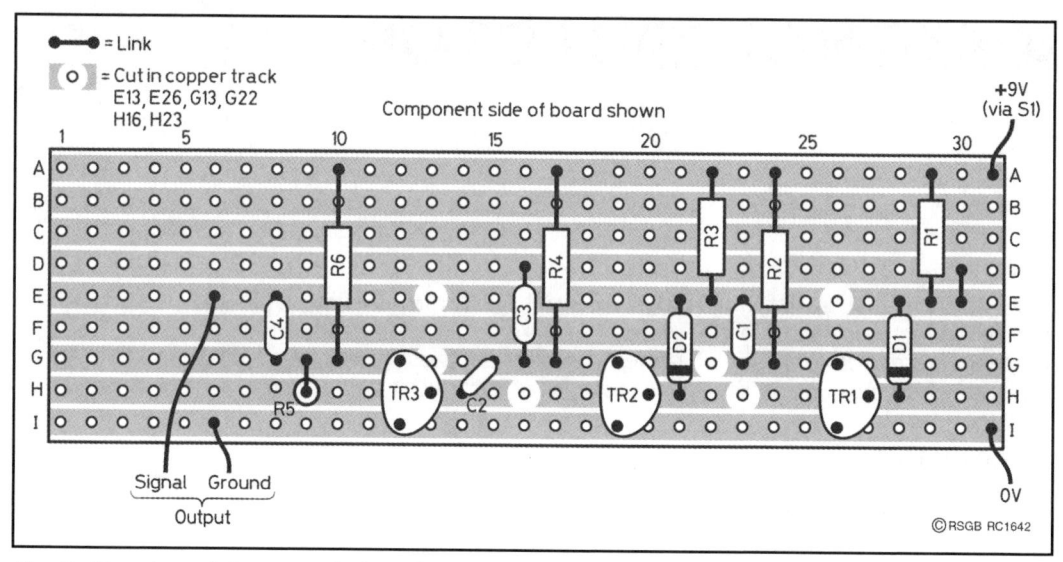

Fig 2. Veroboard layout of the simple signal injector.

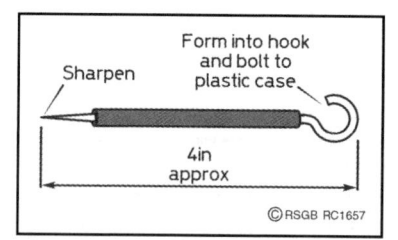

Fig 3. The output probe is made from stiff copper wire. Electrical connection is via a solder tag.

output probe (Fig 3) is made from stiff copper wire, with the insulation left on for the majority of its length and the tip sharpened with a file to a point to ensure good contact. A lead with a crocodile clip is attached to the earth or chassis of the equipment under test.

TESTING AND USE OF THE SIGNAL INJECTOR

To test the circuit, touch the probe to either the input of an audio amplifier or the aerial socket of a receiver. If everything is working, you will hear a buzzing sound.

The basic idea of using a signal injector is to 'chase' backwards through a receiver, listening for the signal. By breaking a receiver into blocks, you can quickly isolate a fault to one block. Once located, you can pursue the fault to component level.

SAFETY NOTICE

The simple signal injector is designed to be connected to equipment which is powered on. Although there is a capacitor to provide protection from the voltages present in working circuits, the output must *not* be connected to any item of equipment which works on voltages higher than 24V. *Television sets or any equipment which uses valves is unsuitable and unsafe for the use of this injector.*

PARTS LIST

Resistors All resistors $1/4$W, 10% tolerance, carbon or metal film
R1, 3   47k
R2, 4   2.2k
R5   8.2k
R6   680

Capacitors
C1–3   10n 16V (minimum) disc ceramic
C4   10n 50V (minimum) disc ceramic

Semiconductors
D1, 2   1N4148 or any silicon signal diode
TR1–3   BC107 / BC108 / BC109 or any similar npn small-signal type

Additional items
S1   On-off switch
Veroboard (stripboard), 0.1in pitch, 9 strips × 30 holes
PP3 battery clip
Hook-up wire
Crocodile clip
Plastic case to suit

AN AUDIO-DRIVEN S-METER FOR DC RECEIVERS

Direct conversion (DC) receivers do not normally have automatic gain control (AGC). The reason for this is primarily that they do not have intermediate frequency (IF) amplifiers, whose gain can be varied to control their output without introducing too much distortion. If the amplifier's output is to be held reasonably constant, then the AGC voltage must track the received signal strength and is therefore a convenient means of driving a signal strength meter.

Alas, the DC receiver does not have this feature. The mixer produces an audio output and is normally directly followed by a low noise pre-amp of fixed gain and wide dynamic range. The primary gain control of the receiver is usually placed after the audio pre-amplifier. With such an arrangement, the audio signal across the gain control is directly proportional to the incoming RF signal over the linear range of the receiver's front end, so why not use the audio output to drive a signal strength meter via a rectifying circuit? The answer to this question is perhaps best appreciated by considering the actual signal strengths typical of the HF bands and how they vary.

S-UNITS

Amateur signal strength reports use the scale 1 to 9, S9 being a noise-flattening solid signal. Way back in the 1940s, commercial receiver manufacturers tried to standardise on a value of 50μV RMS (into a 50Ω load) as being an S9 signal and this is still a good figure to use. When listening on a receiver with an S-meter calibrated to this standard and not having excessively narrow passband, the sound of the signals seem to match the indication, so let's use 50μV as the S9 level and consider what lower levels mean.

One S-point is taken to mean a 6dB change in signal strength, which is a clearly discernible level change to the ear. If the signal is getting weaker, then one S-point down is -6dB. A -6dB change occurs when the signal strength is halved in value, ie reducing from 50μV to 25μV is a one S-point drop. If you continue to halve the signal level eight times you arrive at 0.19μV as equivalent to S1, which is clearly a very weak signal. In fact, constructing an HF receiver which will resolve an S1 signal is no mean achievement.

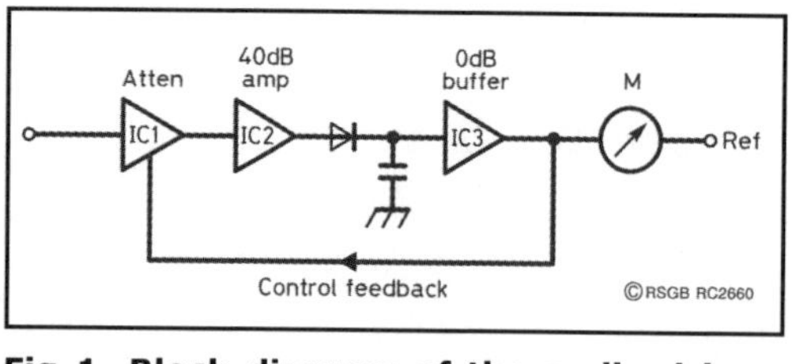

Fig 1. Block diagram of the audio-driven S-meter.

Having established the signal range, if you attempt to display the signal strength

AN AUDIO-DRIVEN S-METER FOR DC RECEIVERS

using a conventional, linear, op-amp style audio rectifier, there is a problem. For example, if you use an ordinary 0 – 10 scaled meter as an indicator with scale digit 9 being S9, then an S8 signal will appear at 4.5 on the scale and S1 would be difficult to read at all. Alternatively, re-scaling the meter to indicate S-points results in a very cramped scale.

Clearly, for an S-meter to be easily read it must display 6dB steps as equal increments. In other words it must have a logarithmic response. This is very difficult to achieve directly in a moving coil meter and is most easily accomplished using non-linear driving circuits. There are several ICs on the market with logarithmic responses, but one of the easiest to use (and the cheapest) is the old Motorola MC3340P. This is a straightforward voltage-controlled attenuator chip and it forms the heart of the following circuit.

USING THE MC3340

The MC3340 is a wideband attenuator chip, claimed by the manufacturers to have a staggering 80dB range. The control characteristic is substantially linear, in dB/V, provided you do not operate at gain levels of 0dB and above. The control voltage range is a little difficult to handle, as it is somewhat dependent upon the supply voltage and has a positive offset of about 2.6V at maximum gain. The manufacturers' performance curves, which I obtained from a data CD ROM issued by Farnell [1] were helpful in establishing the general shape of the IC's response, but the optimum working point was found, in time-honoured fashion, by painstaking measurement. Motorola's ICs do, however, seem to maintain their characteristics from sample to sample; the two I tried were almost identical in performance, even though they were purchased some 15 years apart!

Fig 1 is a block diagram of the meter driver. It is designed to take audio from across the AF gain control and the S9 level is assumed to be 400mV RMS. The input resistance of the attenuator chip, IC1, is about 20kΩ and may need to be buffered if the AF signal comes from a source resistance much above 5kΩ. IC1 is supported by an AF amplifier, IC2, a diode peak rectifier, and a buffer stage, IC3. It was found that for best linearity, IC1 must be operated in a region where it introduces a significant loss of signal and the 40dB amplifier stage IC2 is included in the loop to recover the audio to a level suitable for rectification. After peak rectification and buffering, the resulting DC signal is returned to IC1's control pin, forming a negative feedback loop. This feedback has a twofold effect. Firstly, it stabilises the output of IC1 and secondly, it helps to linearise the control characteristic.

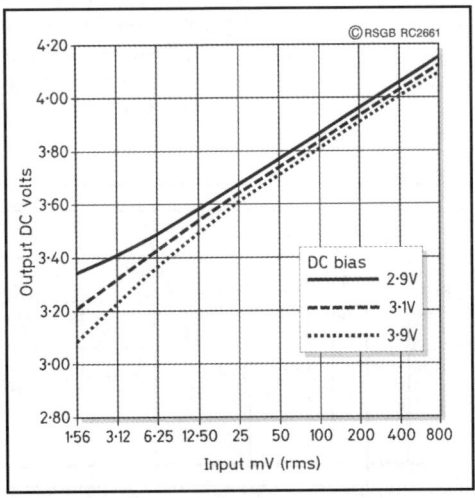

Fig 2 shows the measured response of the circuit, based upon an S9 audio level of

Fig 2. MC3340 output for various bias voltages.

WEEKEND PROJECTS

400mV RMS. The three traces illustrate the effect of changing the zero signal bias voltage. Best overall linearity is achieved with a zero signal bias set at around 3.15V when using a 12V supply. Note the equal increments in control voltage for each doubling of the signal strength.

Thus, if the indicating meter has its negative pole set at 3.15V, its response can be scaled using a series resistor to show S-units directly. A linear increment range of ten S-points can be displayed with this arrangement, which makes the circuit well suited to drive a standard 0 – 10 scaled meter. Again, scale 9 is set as S9 and is achieved with an audio signal input of 400mV RMS. However, this limits the maximum indicated signal to one S-point over S9. Compared to a commercially-produced receiver, with an S-meter scaled to 60dB over S9, this performance may look limited until you realise that such an indication means a signal 1000 times more powerful than S9 and is perhaps of limited application! Practically speaking, since this circuit is intended to be added to DC-type receivers, which usually employ an RF attenuator to cope with extra-large signals, if steps that are multiples of 6dB are used, the range can be extended without the need to modify the meter scale. An RF attenuation of 6dB simply adds one S-point to the meter indication, and so on.

GENERAL CIRCUIT POINTS

The final circuit is shown in Fig 3 and PCB layout in Fig 4. Stable DC supplies are essential. A 12V regulated supply was available in my DC receiver and I used that in conjunction with a low power regulator, IC4 (78L05), to fix the 5V bias rail voltage. The circuit will work just as well from a 15V regulated supply, but IC1's bias voltage would need to be adjusted to 4V. Operation from lower supply voltages is not recommended, as linearity suffers.

The meter used should have a full scale sensitivity in the range of 100 to 400µA. Small meters with light pointers are the best types to use as they can follow variations in signal strength quickly, the movement having low inertia. Good VU indicators meet this requirement but are usually scaled in 3dB steps and need rescaling. Helpfully, because the meter is driven from the output of an op-amp, the actual full scale sensitivity is of little importance - provided you do not exceed the

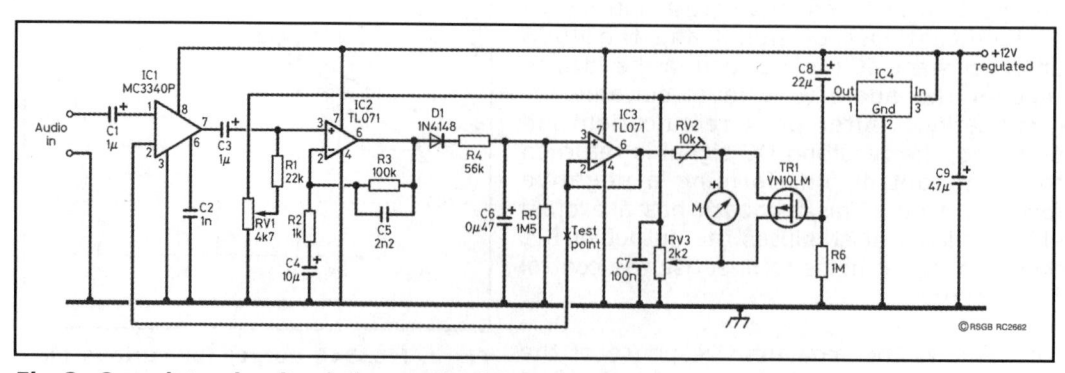

Fig 3. Complete circuit of the audio-driven S-meter.

AN AUDIO-DRIVEN S-METER FOR DC RECEIVERS

maximum current output of the amplifier. The value of RV2 may be varied to suit the meter if outside the range specified.

To overcome switch-on transients, which tend to cause vigorous full-scale pointer movement, C8 is used to bias the MOSFET, TR1, temporarily into conduction and short out the meter. After a second or two, when the circuit capacitors have reached their working voltages, C8 has also charged via R6 and the circuit reverts to its normal state as TR1 switches off.

INITIAL ADJUSTMENT

Setting up the circuit is quite straightforward, but does need a DC meter and an audio signal generator. Firstly, short the audio input to guarantee no signal. Adjust RV2 and RV3 to mid-range. Connect the DC meter to the test point or the output of IC3 (pin 6). Now switch on the supply and adjust RV1 for the correct zero signal bias, appropriate to the supply voltage you are using. Next, set RV3 for a zero reading on the S-meter. Finally, apply 400mV RMS at about 800Hz to the audio input and adjust RV2 for an S9 reading. Repeat this cycle of events until stability is reached. Following this routine, it is comforting to reduce the AF signal input to half its level and see that the meter indication reduces by one S-point. It is possible to repeat this action right down to S1, but beware, you need a well earthed/screened test bench when the input signal level is down to a few millivolts.

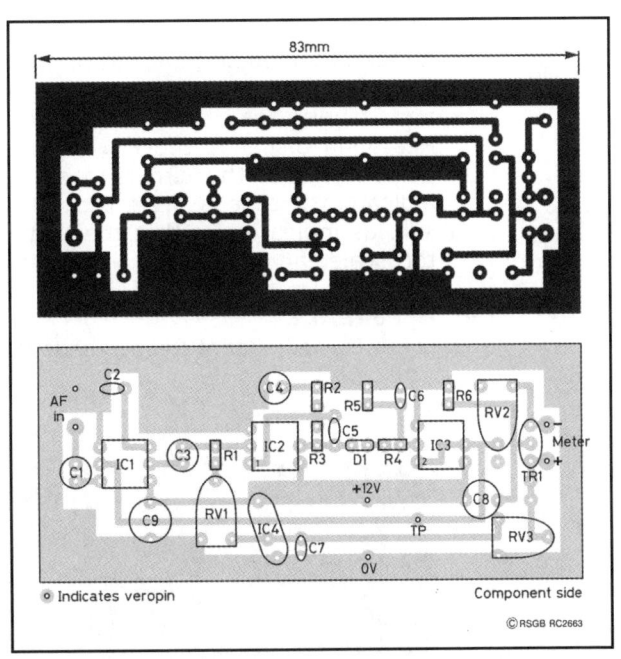

Fig 4. PCB layout and component placement.

Connecting the meter to the receiver audio requires that an audio level of 400mV be produced across the AF gain control for a 50µV input signal. This may require a low gain amplifier, either to achieve the audio level or to prevent loading of the audio stage in the receiver. Fig 5 is suggested if such an amplifier is needed. Remember, the input resistance of the MC3340P is about 20kΩ. Unless this is high compared to the output resistance of the receiver audio pre-amp stage, it may load the amplifier and perhaps cause distortion. Also, if this pre-amp is used, the polarity of the input capacitor C1 should be reversed. This is because the pre-amp output is at +5V.

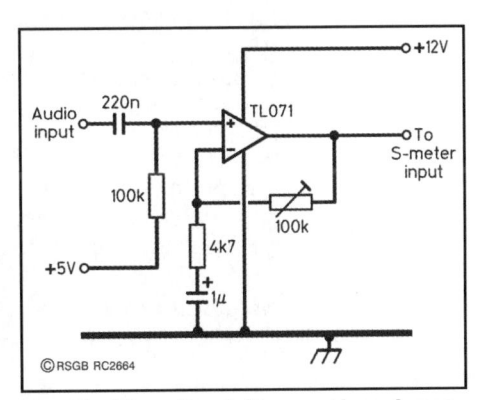

Fig 5. Circuit of the optional pre-amp.

143

WEEKEND PROJECTS

At G3DXZ, the current low-band DC receiver sports an edgewise S-meter, obtained surplus and re-scaled. With many of this type of meter, the scale is a strip of paper that simply clips into the plastic housing and is very easy to remove, lay flat and re-scale or simply replace. Rescaling also has the advantage of allowing you to shift the 10 S-point range of the circuit to start, say, at S3 and give some over-9 indication if that is preferred. The real bonus in using one of these surplus indicators is that they often have a built in lamp to illuminate the scale, and that does add a touch of class!

REFERENCE
[1]   Farnell, Canal Road, Leeds LS12 2TU. Tel: 0113 263 6311.

PARTS LIST
Resistors (all fixed resistors, 0.25W 5%)
R1 22k
R2 1k
R3 100k
R4 56k
R5 1.5M
R6 1M
RV1 4.7k mini carbon pre-set
RV2 10k mini carbon pre-set
RV3 2.2k mini carbon pre-set

Capacitors
C1 1µF 63V electrolytic
C2 1nF 63V mylar
C3 1µF 63V electrolytic
C4 10µF 63V electrolytic
C5 2.2nF 63V mylar
C6 0.47µF 63V electrolytic
C7 100nF 63V mylar
C8 22µF 16V electrolytic
C9 47µF 16V electrolytic

Semiconductors
IC1 MC3340P
IC2 TL071
IC3 TL071
IC4 78L05
TR1 VN10LM
D1 1N4148

Miscellaneous
M1 100 – 400µA

A TIME-OUT UNIT FOR DIGITAL MODES

The author has run digital modes for a couple of years and, after a number of close calls, has designed a simple time-out circuit around the 555 timer IC. The original setup used a transistor/relay arrangement to provide isolation between the computer and the transceiver PTT line and, for convenience, the timer circuit has been added.

The article 'Computer-to-Radio Interfaces', elsewhere in this book, covers isolation of computers and transceivers. There is always, however, the possibility of the program or computer crashing with the PTT line remaining in the transmit state. This may not be too problematic in SSB mode if there is no modulation or if the operator is at the keyboard. But in Packet/*APRS*/*UI-View* modes, with FM modulation, and where the operators are, more often than not, absent for prolonged periods, the effect on the transmitter output stage can be catastrophic.

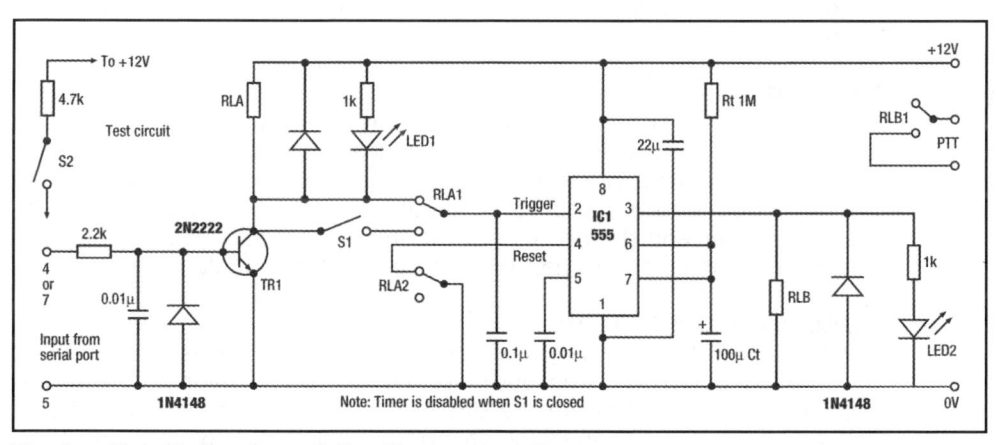

Fig 1. Circuit diagram of the time-out unit.

CIRCUIT DESCRIPTION
When the serial port goes positive (Fig 1) the transistor TR1 conducts, pulling the collector down to near 0V, energising RLA. During the very short period before the RLA contacts actually make, the trigger terminal of the IC, pin 2, is also pulled down via the relay contacts RLA1, triggering the 555 timer period.

WEEKEND PROJECTS

The output of the IC, pin 3, now goes to almost 12V, energising the relay, RLB, and closing the PTT circuit via the contacts, RLB1. The timer resistor Rt will now start charging the timing capacitor, Ct, and when the voltage at pin 6 of the IC reaches an internal IC reference voltage, the output on pin 3 will fall to 0V releasing relay RLB and breaking the PTT circuit. When the serial port returns to 0V, relay RLA will be released bringing the reset terminal of the IC, pin 4, to 0V by way of contacts RLA2, which resets the timer. Similarly, if the serial port returns to 0V before the timing period has expired, RLA will be released, the timer will be reset, releasing RLB and breaking the PTT circuit.

Summarising, when the serial port goes positive, the transmitter is keyed until such time as the serial port returns to 0V or the timing period, set by Rt and Ct, has expired. The timer period is controlled by Rt and Ct, according to the expression Tp = 1·1 x Rt x Ct seconds. With the values shown in the circuit, the period should be about 110 seconds. In fact, the breadboard and two working models were both in the region of 120 seconds, no doubt due to tolerances of the capacitors. The timing period can be modified by the adjustment of Rt and Ct, but increasing Rt beyond 1MΩ is not recommended.

Beware of leaky capacitors that will have a significant effect on the timer period with such values of resistor. The LEDs are optional. LED1 indicates that the computer is signalling a 'transmit' condition and LED2 indicates that the PTT is activated. The switch, S1, when closed, turns the timer function off by holding the trigger terminal in a continuously-triggered state when the serial port is positive, ie signalling a 'transmit' state.

TESTING
The timeout period can be checked by connecting the input to the 12V supply via a 4.7kΩ resistor, such that it presents about 5V to the input terminal. When S2 is closed, both LEDs should glow and when the timing period expires, and LED2 should extinguish. Opening S2 will reset the timer and the process can then be repeated. If S2 is opened before the timing period has expired, both LEDs should extinguish.

DUAL-PORT OPERATION
Depending on the software, it is possible to run two separate programs on one soundcard by using both the left and right channels. The left channel being presented on the tips of the 'line in' and 'line out' stereo jacks and the right channel on the rings. You will probably find that microphone sockets are mono and, where dual-channel operation is required, it will be necessary to use a 'line' or 'auxiliary' input.

Pin 7 of the serial port is the left channel and pin 4 is the right. Pin 5 is common for both channels.

EARTHING
It is desirable to isolate the computer completely from the radio equipment. If this cannot be done, it may be preferable to bond down all equipment to the station earth.

A TIME-OUT UNIT FOR DIGITAL MODES

PARTS LIST
Resistors – all 0.25W
1 off 2.2kΩ
2 off 1kΩ
1 off 1MΩ
1 off 4.7kΩ. For testing only

Capacitors – all 16V minimum
2 off 0.01μF
1 off 0.1μF
1 off 22μF
1 off 100μF

Semiconductors
1 off 2N2222 npn transistor
3 off 1N4148 diode
1 off NE555N integrated circuit
2 off standard LED, any colour

Relays
2 off   miniature DPDT relay, 700Ω coil, 12V, Maplin UQ89W

Miscellaneous
2 off   SPST switch. One is required but only for testing

A T-match ATU

An aerial tuning unit (ATU) is very useful if you are a licensed radio amateur or a short-wave listener. The purpose of an ATU is to adjust the aerial feed impedance so that it is very close to the 50Ω impedance of the receiver or transmitter, a process known as *matching*. When used with a receiver, an ATU can dramatically improve the signal-to-noise ratio of the received signal. On transmit, the aerial must be matched to the transmitter so that the power amplifier operates efficiently.

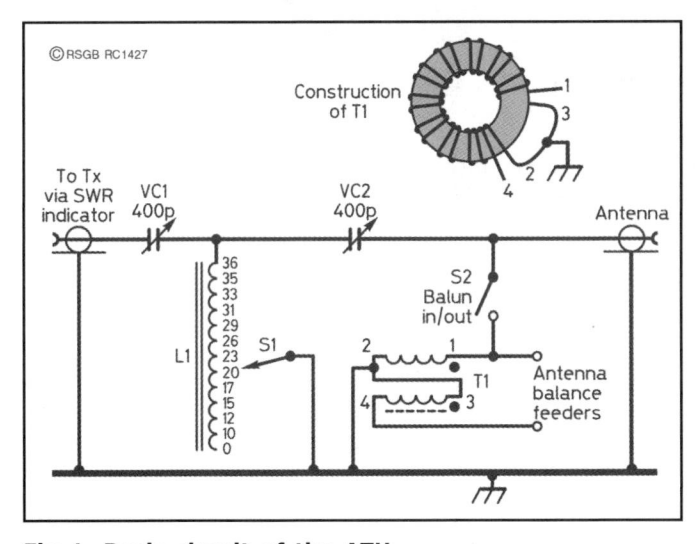

Fig 1. Basic circuit of the ATU.

DESIGN

The basic ATU design is called a *T-match*; you can see the basic shape of the letter 'T' reflected in the layout formed by the components VC1, VC2 and L1 in Fig 1. The circuit will match the coaxial output of the transceiver to an end-fed aerial or to a coaxial cable feed to the aerial. This design also uses a *balun* (*bal*ance to *un*balance) transformer for use with aerials using twin-wire balanced feeder.

This unit will handle up to 5W and operates over the frequency range of 1.8 to 30MHz.

CONSTRUCTION

Inductor L1 is wound on a T-130-2 powdered-iron toroid. The inductor is tapped and fixed to the tags of a 12-way rotary switch. Taps are formed by making a loop about 1cm long in the wire and twisting it tightly. The loops are scraped clean of enamel and tinned with solder ready to be soldered to the tags of the 12-way rotary switch. If the loops are about 1cm long, it is just possible to bend them to fit the tabs of the switch without having to extend them with short wires.

The balun is wound with two wires twisted together, a process known as *bifilar winding*. These two wires can be twisted together (before winding on to the toroid) by fixing one pair of ends in a vice, the other

A T-match ATU

ends in a small hand-drill. The drill is then slowly rotated so that the two wires are twisted together neatly. Identify the start and finish of each winding with a buzzer and battery or an ohmmeter. The finish of the first winding is joined to the start of the second, as shown in Fig 1. The two capacitors are twin 200pF polyvaricon capacitors with both gangs connected in parallel to give 400pF max.

You have to drill holes in the box for the control shafts of the capacitors and the switch, and the RF sockets. The switch is fixed using a nut on the control shaft and the two capacitors are fixed using adhesive (hot melt glue is preferred). Take care not to let any tags from the capacitors touch the box as both sides of both capacitors are not earthed.

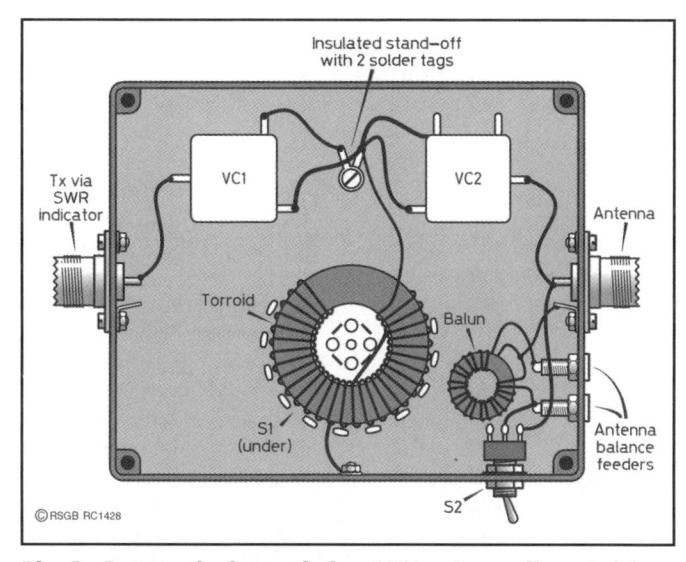

Fig 2. Internal view of the ATU. A small metal box makes a suitable housing.

An appropriate RF socket, such as a SO-239, BNC or phono socket may be used – the choice is yours and should suit your existing equipment. 2 – 4mm sockets may be used for the balanced output. See Fig 2 for the layout, and the photograph for the finished product. It is a good idea to make a graduated dial for each of the three control knobs. An alternative is to use calibrated knobs. You can then calibrate the settings of the three controls so that they can rapidly be reset when you change frequency bands.

The completed ATU.

OPERATION

The best indication of optimum matching can be achieved using a VSWR bridge; the ATU controls are adjusted sequentially and several times for minimum VSWR. If used for receive only, the best aerial-to-receiver match can be achieved by adjusting the controls for maximum signal.

PARTS LIST
Capacitors
C1, 2   200 × 200pF

Inductors
L1   T-130-2 powdered iron toroid with total 36 turns of 22SWG enamelled copper wire, tapped at 10, 12, 15, 17, 20, 23, 26, 29, 31, 33 and 35 turns from the earth end
T1   12 bifilar turns of 26SWG enameled copper on FT-50-43 toroid

Additional items
S1   1-pole 12-way rotary switch
S2   SPST toggle switch
RF connectors, SO239 sockets or similar (see text)
Aluminium or die-cast metal box
Three plastic knobs for capacitors and switch
Two 4mm sockets for balanced aerial connection

A 1750Hz TONEBURST FOR REPEATER ACCESS

Repeaters across the UK and much of Europe need an access tone to switch the transmitter from standby ready for use. Commonly, this is a 1750Hz tone of duration no greater than half a second. Although many repeaters may now be accessed using the continuous tone-coded squelch system (CTCSS – see the *RSGB Yearbook*), you may wish to access a repeater whose access frequency you don't know; in this case, using the universal 1750Hz tone will gain you access. Commercial transceivers are fitted with an automatic toneburst, but if you are using a home-made design, then you may want to incorporate this little circuit.

Warning – CMOS!

This circuit uses a member of the integrated circuit family known as CMOS (complementary metal-oxide semiconductor). These use very little current and can be completely destroyed if they come into contact with the magnitudes of static electricity that most of us carry about when we walk on carpets and wear rubber shoes. You will never know if this wanton destruction has happened – all you will discover is that your circuit doesn't work and that you have tested everything. To avoid this problem do the following things:

1. Before you open the little packet in which the IC is supplied, touch something which you know to be earthed – the metalwork of any equipment which is mains earthed, for example. Then open the packet.
2. Let the IC fall gently on the bench – don't pick it out with your fingers. Touch your earthed metalwork again. Pick up the IC and insert it gently into its holder.

The circuit is safe from destruction while it is connected to the battery.

CIRCUIT DESCRIPTION

The circuit is shown in Fig 1. The tone is generated by an integrated circuit oscillator (IC1), whose frequency is controlled by a ceramic resonator, XL1. Its frequency is very high, and is divided down to the 1750Hz needed by the same chip.

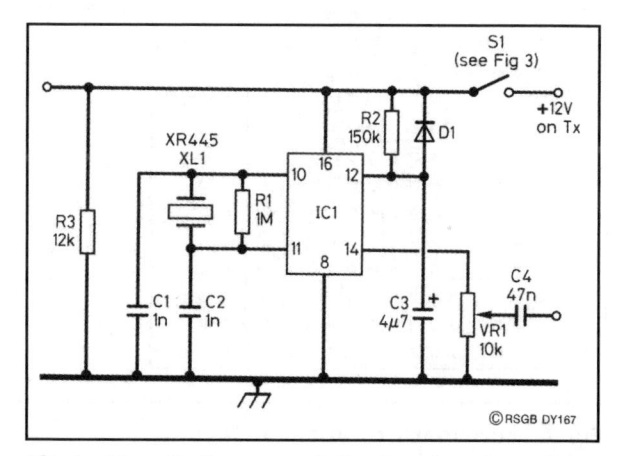

Fig 1. Circuit diagram of the toneburst module.

WEEKEND PROJECTS

The ceramic resonator is designed to operate at 455kHz, the intermediate frequency of many receivers. Because all divider circuits use powers of 2, we need the oscillator to run at 448kHz so that when it is divided by 256 ($256 = 2^8$), we end up with 1750Hz. Try it on your calculator:

448000 / 256 = 1750.

We use C1, C2 and R1 to pull the frequency of the oscillator away from 455kHz to 448kHz. The divider chain has eight counters in it, and each counter divides the frequency of the signal it sees by two, giving the final division of 256.

This counting process can be stopped at any time by taking the voltage on the reset pin (pin 12) up to the supply voltage. When the circuit is switched on by closing S1, pin 12 is at 0V because C3 is discharged. The oscillator runs, producing the output frequency of 1750Hz. As time progresses, C3 charges up through R2 and the voltage on pin 12 rises. When this has risen sufficiently, and in a time determined by the values of C3 and R2, the counter resets and stays in the reset state; no division takes place and there is no output. The duration of the toneburst is thus governed by C3 and R1.

When S1 is opened, the circuit is switched off, and C3 is discharged through D1 and R3, ready for the next toneburst. If you have used a repeater, you will know that a toneburst is needed only to activate a repeater in the standby condition; it is not needed once a contact has been established.

VR1 adjusts the amplitude of the tone fed to the microphone, and C4 prevents any voltage that may be present on your microphone connector from damaging the integrated circuit.

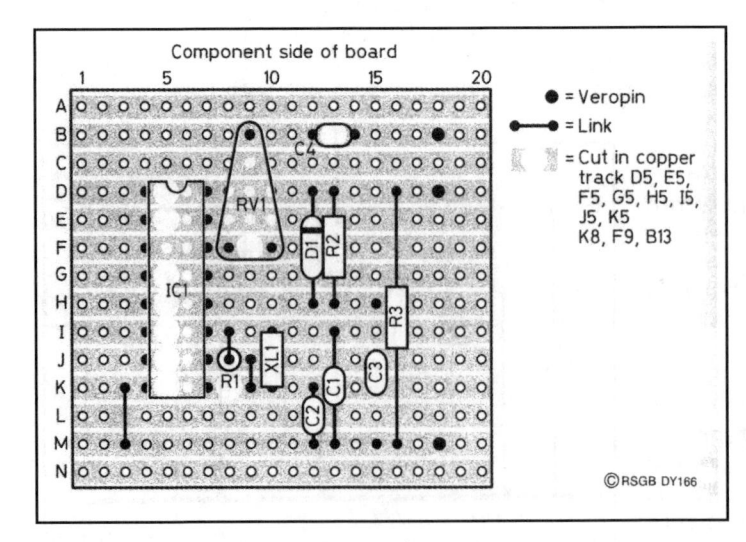

Fig 2. Veroboard component layout.

CONSTRUCTION

The prototype circuit was built on Veroboard of the copper-strip variety, measuring 20 holes by 14 strips. The layout is shown in Fig 2. Make the track cuts first, and check that there are no slivers of copper wedged between adjacent tracks. Then, solder in the IC socket (with the notched end facing towards track A), the wire links and the three Veropins. Having done this, solder in the resistors, capacitors and diode, making sure that D1 and C3 are the right way

A 1750Hz TONEBURST FOR REPEATER ACCESS

round! Using your best soldering technique, solder in the ceramic resonator quickly, to prevent heat damage. Recheck your circuit, check for solder splashes and bridges, and then gently insert IC1 into its socket, matching up its notch with that of the socket.

TESTING
Set VR1 to half-way and connect a crystal earpiece to the output; apply power to the circuit. You should hear the tone, lasting for about half a second. If there is no tone, disconnect your circuit from the power supply, and check for dry joints in the vicinity of pin 12. Is the diode, D1, the correct way round? Is C3 the correct way round? Did you choose to ignore the CMOS safety precautions given earlier?

Once the circuit is working, you need to decide how you are going to connect it to your transmitter. Two options are shown in Fig 3. If you have access to a point in your transmitter circuit that has between 9V and 12V positive on it during transmit, you can use this to power your circuit. As the toneburst is needed only for repeaters, the switch, S1, disconnects it when not needed, as shown in Fig 3(a). 1f you want the circuit to be self-powered, then a 9V PP3 battery may be used; Fig 3(b) shows this configuration.

The output from the circuit board is fed directly into the microphone socket, in parallel with the microphone itself; use thin coaxial or screened cable for this lead, or you may induce hum into the microphone circuit and suffer from RF breakthrough into the audio circuits. To adjust the setting of VR1, start with it at the zero output position and connect a dummy load to your transmitter. Slowly, increase the output while monitoring your transmitted signal on another nearby receiver. Make sure you do not increase the output so far that the signal sounds distorted. If you would prefer that the tone was on continuously while you made this adjustment, simply connect a wire across C3 remembering, of course, to remove it as soon as you have completed the test!

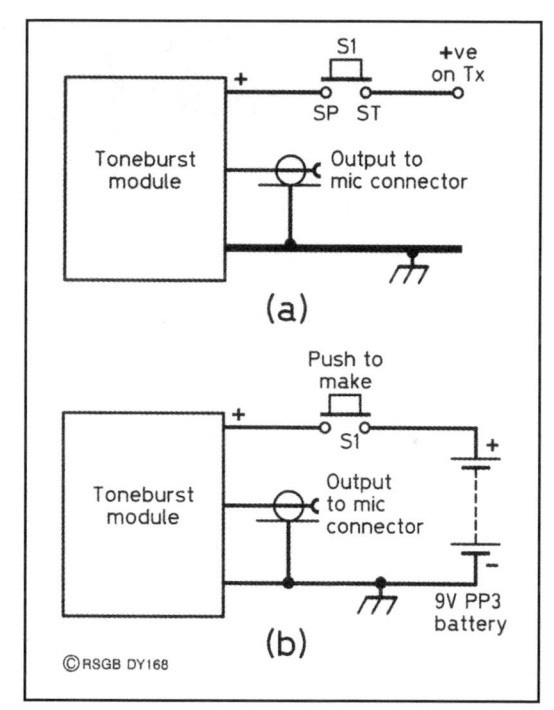

Fig 3. Alternative switching arrangements.

WEEKEND PROJECTS

PARTS LIST
Resistors: all 0.25W, 10% tolerance (or better)
R1   1 megohm (MΩ)
R2   150 kilohm (kΩ)
R3   12 kilohm (kΩ)
VR1   10 kilohm (kΩ) horizontal preset

Capacitors: all 16V wkg or higher
C1, C2 1 nanofarad (nF) disc ceramic
C3   4.7 microfarad (μF) tantalum bead
C4   47 nanofarads (nF) disc ceramic

Semiconductors
IC1   4060
D1   1N4148

Additional items
XL1   XR4SS
        Veroboard (see text for size)
        Veropins (3)
S1   Switch (momentary action, push-to-make, SPST)
        16-pin DIL socket for IC1
        Single-core insulated wire for links
        Coaxial or screened cable for microphone connection

A COLOURFUL VOLTAGE MONITOR

Probably the most common unit used in electrical, electronic and radio engineering is the volt, which is a measure of the amount of energy associated with the electrical charge of electrons at a particular point. One volt is defined as one joule of energy per coulomb of charge, and there are two main situations where it is encountered. The first is where a voltage is 'dropped', for example across a resistor, and this is a measure of the energy lost as a current flows (ie electrons move) against that resistance. This is usually referred to as *potential difference* (PD). The other is where a device 'pushes' the electrons, in the form of an electrical current, around a circuit, as does a signal generator or a battery. In this case the voltage is usually referred to as the *electromotive force* (EMF). So when measuring the voltage of a battery we are measuring how much energy the battery is capable of giving to the circuit of which it is part. As the energy stored in the battery is used up, its output voltage falls, an indication of the falling amount of energy available to 'drive' the electrons. Often the degree to which a battery has been 'used up' is of more interest than the actual voltage output, and in this case an indication of a voltage 'range' is more useful.

WHAT IT DOES

This project uses a series of LEDs to indicate when a voltage falls below 12V, when it is inside an acceptable range of 12 to 13V, when it is inside a second acceptable range of 13 to 14V, when it is a little too high at between 14 and 15V, and to warn when it is above a maximum acceptable 15V. With these ranges, it is ideal for monitoring car battery voltage, and when using rechargeable batteries to power equipment around the shack or 'in the field'.

The completed unit. The components are laid out on stripboard. The monitor could be put into a project box or the LEDs mounted separately if desired.

HOW IT WORKS

To give a meaningful output in the form of a sequence of ranges, a series of reference points is needed with which to compare the test

WEEKEND PROJECTS

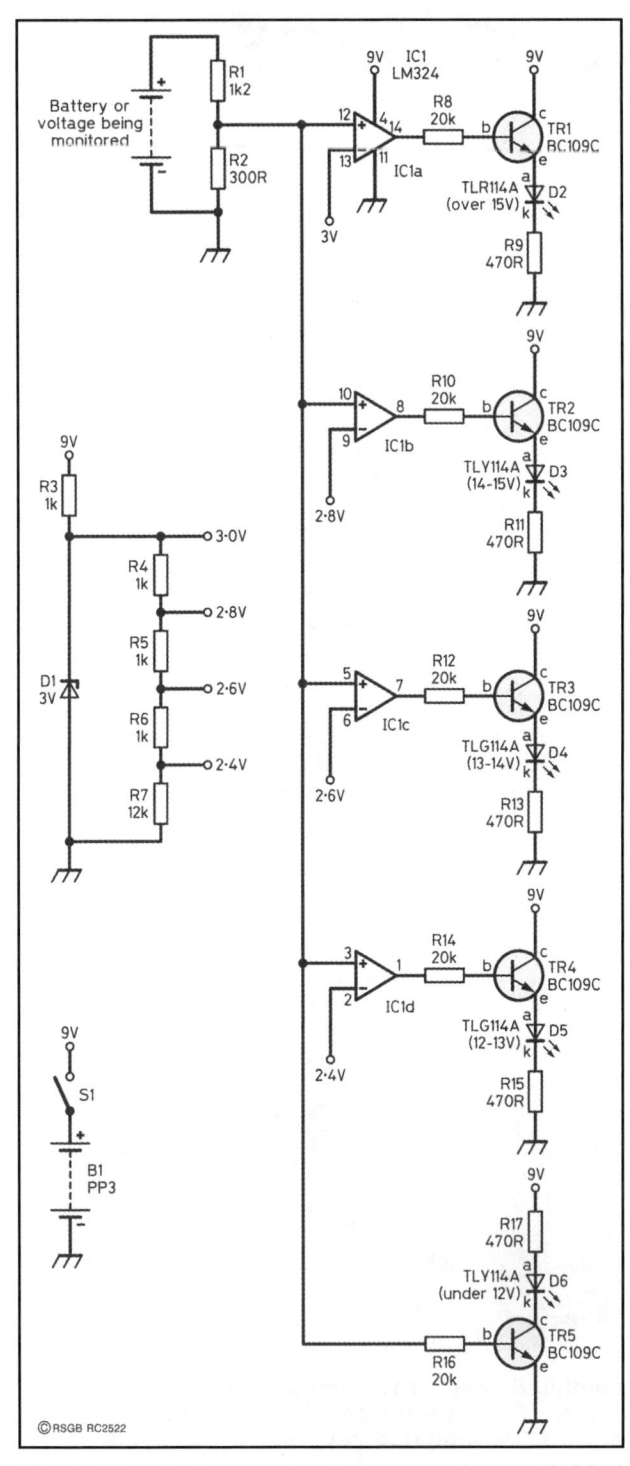

Fig 1. The device works by comparing a divided fixed voltage to a divided sample of the voltage being monitored.

voltage. These reference voltages can easily be generated using a resistance divider chain. However, the main reference from which all these are derived needs to be stable at all times, and it is not sufficient simply to connect the resistor chain across the supply battery. If this technique were to be used the reference voltage would of course change as the supply battery aged or when it was subject to different load currents. In this project a Zener diode is used to achieve a stable reference voltage. These devices are semiconductor diodes connected 'backwards', ie reverse-biased and, when used like this, the normal flow of current through them is blocked. However, all semiconductor devices have a 'leakage' current which is a small current that passes in the 'wrong' direction, and Zener diodes are designed to give a particular fixed voltage across them when this leakage current is flowing.

Each reference voltage from the resistance divider chain is fed to the inverting input of its own comparator, the test voltage being connected to all of the non-inverting inputs. Thus the output of an individual comparator will be around 0V unless the test voltage rises above the comparator's reference voltage, at which point it will switch over to a value close to the positive supply voltage. By connecting each output to an LED with its own driver transistor to provide enough current, the LEDs will light up in turn as the test voltage increases.

Since the test voltage will normally be higher than the circuit supply voltage, the test voltage

A COLOURFUL VOLTAGE MONITOR

and the reference voltages are both scaled down by the same amount, in this case by a factor of five. The test voltage is scaled by a simple voltage divider, the reference voltages by using a 3V Zener diode (the maximum reference voltage before scaling being 15V).

The '<12V' indication is given by switching the LED driver directly from the scaled test voltage. In this way a false '<12V' indication is not given when there is no test voltage connected.

CIRCUIT

The circuit diagram for the voltage monitor is shown in Fig 1. The comparator functions are obtained by using the operational amplifiers (op-amps) of IC1 wired without any feedback. In this configuration the op-amp's high gain means the output can only ever be equal to the positive or negative (0V) supply voltage, depending upon whether the voltage at the non-inverting input is more positive or more negative than the voltage at the inverting input.

The transistors all operate as switches, ie they are either 'on' (conducting) or 'off' (non-conducting). When they are 'on', the transistor passes enough current to light the LED (the op-amp output alone cannot reliably provide enough current to do this). The values of resistors chosen in the transistor / LED part of the circuit are those that limit voltages and current flows to the working values of the transistors and LEDs.

CONSTRUCTION

The stripboard layout for the project is shown in Fig 2. The LM324 quad op-amp integrated circuit, the

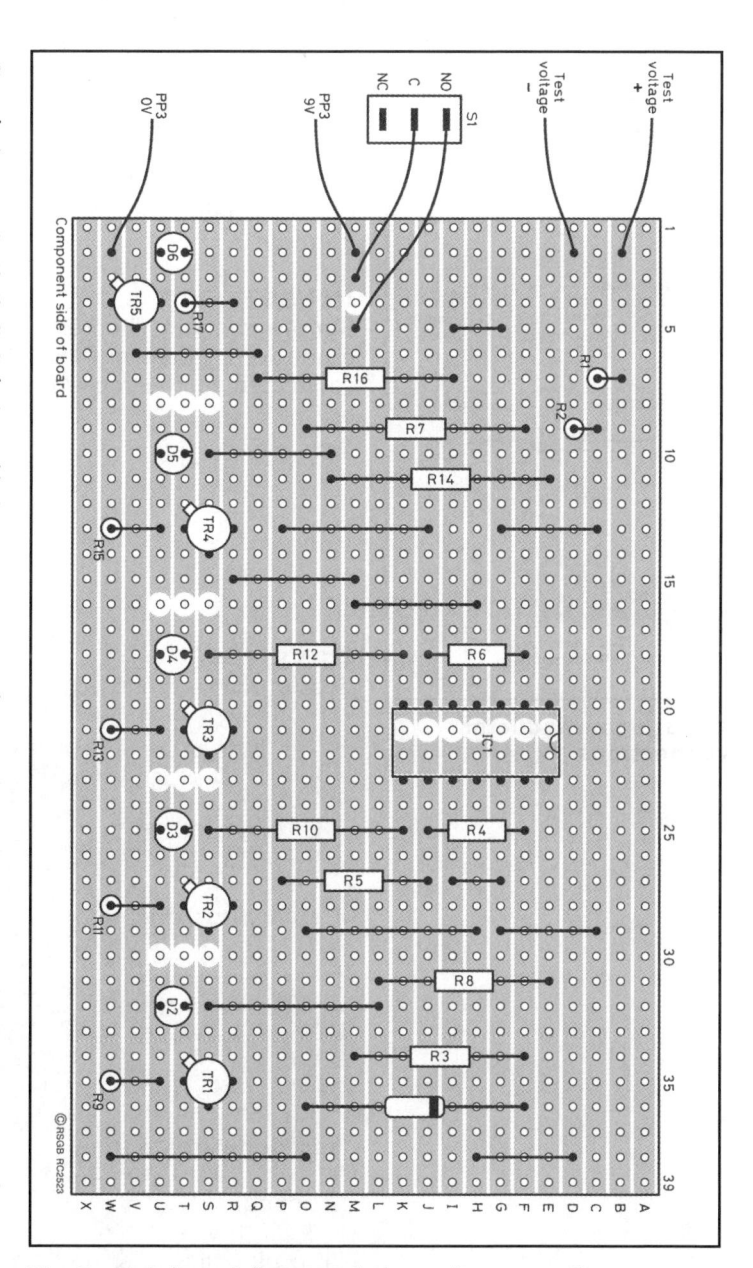

Fig 2. Stripboard layout of the voltage monitor.

157

WEEKEND PROJECTS

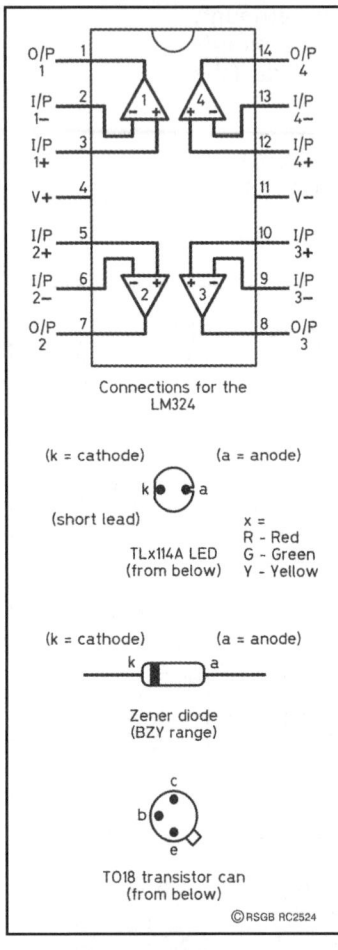

Fig 3. Semiconductor connections.

transistors and the LEDs all need to be connected the 'right way round', and Fig 3 shows the correct orientation of these devices. An important point to note is that the 'positive' test lead needs to be connected to the more positive terminal of the battery under test, so it is useful to use red and black wires respectively for the test leads.

COMPONENTS
An LM324 quad op-amp was chosen as it gives four devices in a single package and can be powered from a single supply voltage. The transistors are a general-purpose npn type, and similarly the LEDs are different colours of a general-purpose type. The lower the tolerance value of the resistors used for the voltage divider chain, the more accurate are the reference voltages.

IN USE
Simply connect the test leads to the voltage being monitored (ensuring that the 'positive' test lead is connected to the more positive terminal), and switch on!

PARTS LIST
Resistors - All resistors metal film, 0.6W 1%
R1	1.2k
R2	300
R3 – 6	1k
R7	12k
R8, 10, 12, 14, 16	20k
R9, 11, 13, 15, 17	470

Semiconductors
IC1	LM324
TR1 – 5	BC109C
D1	BZYC3 Zener, 3V 0.5W
D2	TLR114A red LED
D3, 6	TLY114A yellow LED
D4, 5	TLG114A green LED

Additional items
S1	SPDT
Stripboard	
PP3 battery and clip	
Crocodile clips	

VOLTAGE REGULATION

Nothing is perfect; the AC mains, transformers, resistors, transistors, integrated circuits – we know how they *ought* to behave but how they *really* behave can be rather different.

THINGS CHANGE...
The AC mains is not constant, particularly if you live out in the country and are served by several miles of overhead cable. Your next-door neighbour switches on her electric oven, and your lights dim! Transformers have losses in the resistance of their windings, and these losses vary depending on the load at any given time. Resistors produce a voltage drop when a current flows through them. The current may cause a resistor to heat up unduly, change its resistance, and produce a different voltage drop. Transistors are also sensitive to changes in temperature, and may cause major problems in a poorly-designed circuit.

Suppose you are building a circuit using TTL logic chips. On the data sheet specifying how they should be used, you see that the TTL family of integrated circuits requires a supply of 5.00 ± 0.25V. Whatever happens in the rest of your equipment, and as a result of any mains variation, must not cause more than 0.25V variation of the 5V supply. We shall now see how we can stabilise our voltage supplies so that they are virtually independent of input voltage variations.

THE ZENER DIODE
For voltages below about 30V, the heart of almost all voltage stabilisation (or regulation) circuits is the Zener diode. This is a discrete silicon device, which has the normal diode characteristic when forward-biased (ie when the anode is positive with respect to the cathode), but when reverse-biased it passes negligible current up to a well-defined point at which the current increases sharply. This is illustrated in Fig 1, which also shows the circuit symbol.

The point at which the reverse current starts to flow can be closely controlled in manufacture between about 3V and 200V. If such a Zener diode is intentionally biased into this steep region, the voltage across the diode varies very little for large changes in current; the device is serving to

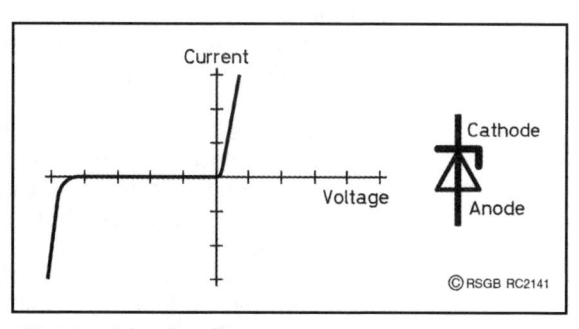

Fig 1. The circuit symbol and characteristic of a typical Zener diode.

WEEKEND PROJECTS

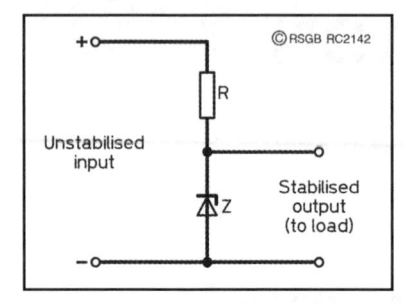

Fig 2. A simple circuit that produces a stabilised output from an unstabilised input.

stabilise (or regulate) the voltage across it. A typical circuit is shown in Fig 2. The value of R is given by the equation

$$R = \frac{V_{out}(V_{in} - V_{out})}{P},$$

where V_{in} is the unstabilised input voltage, V_{out} is the stabilised output voltage (the Zener voltage), and P is the Zener dissipation in watts (see text).

Zener diodes are specified by two parameters – the Zener voltage and the maximum power dissipation of the device. For example, all Zener diodes of the types BZY88C and BZX55C have 0.5W dissipation and are by far the commonest. A 15V Zener would be marked BZY88C15V; you are expected to know it is a 0.5W device!

SEE-SAW ACTION

A see-saw is a simplistic, but very useful, analogy of how the Zener works in a circuit. In Fig 2, the current from the unstabilised supply has two paths – through R and Z to earth, or through R and the load connected across Z. Suppose we have chosen a 10V Zener of the BZY88C series. It is a 0.5W device and therefore it will draw a current I given by:

$$I = \frac{P}{V} = \frac{0.5}{10} = 0.05A \text{ or } 50mA.$$

Suppose no load is connected. All 50mA will go through the Zener as there is no other route. Suppose the load now takes 5mA; here the current see-saw comes in. With 5mA going through the load, 45mA now goes through the Zener. If the load takes 10mA, then 40mA flows through the Zener. This keeps the total current constant (from the unstabilised supply), and thus helps to keep the voltages constant also.

This see-saw action is not perfect, particularly as the load takes more and more current. As the current through the Zener approaches zero, the see-saw action fails and the stabilisation becomes inoperative. This circuit is simple and works well, provided the demands of the load are small, but it has one big disadvantage – your power supply is delivering the maximum current at all times, whether or not the load takes any. Notice in the example that the Zener takes 50mA when there is no load. Notice also that the unstabilised voltage must always be significantly greater than the stabilised voltage because of the voltage drop across R. It is inadvisable to try to construct a 12V stabilised supply from a 13V unstabilised input! As the range of currents demanded by the load increases, so also should the difference between the supply voltage and the regulated voltage.

VOLTAGE REGULATION

A REFERENCE VOLTAGE

Removing the need for the power supply to supply the maximum power at all times can be achieved by keeping the current through the Zener constant. In this way, it does not need to dissipate maximum power and the regulation it provides is much better. It is used more as a reference voltage than as a type of voltage source. This idea is illustrated in Fig 3. The stabilised voltage produced by the Zener diode is applied to the base of the transistor, which acts as an emitter follower, producing about 0.6V less than the voltage across the Zener but at a current limited only by the capacity of the power supply and the characteristics of the series (or 'pass') transistor TR1 which usually requires a heatsink. There is a shunt-stabilising version of this circuit but it is little used because of its inefficiency.

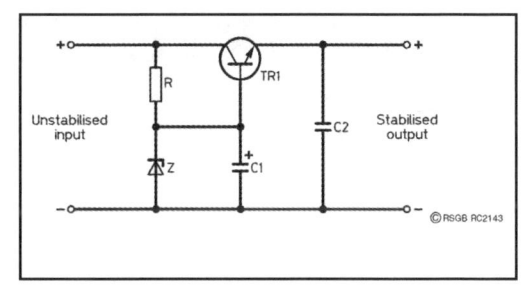

Fig 3. An improved circuit.

To take this circuit to its logical conclusion and produce a well-stabilised supply, the circuit of Fig 4 is common. This produces, by means of the integrated circuit op-amp, a very sensitive comparison between the Zener voltage, V_z, and the output voltage divided down by R2 and R3. This allows the output voltage to be greater than the Zener voltage and, depending upon the voltages and currents involved, can obviate the need for extra smoothing capacitors. Readers interested in refinements such as foldback current limiting and crowbar protection are referred to references [1] and [2]. Needless to say, integrated circuits have been produced that possess all the qualities (and some refinements, too) discussed so far but are for the lower output current ranges only.

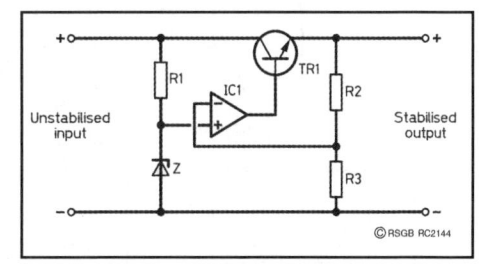

Fig 4. Using an operational amplifier to improve the circuit even further.

These are typified by fixed positive-voltage devices such as the LM78**CT series of 1A regulators and their negative-voltage equivalents, the LM79**CT series. The LM338K is a 5A regulator for output voltages between 1.2V and 32V. The LM317T is a popular 1.5A variable regulator.

HIGHER VOLTAGES

When stabilisation is required at high voltages, solid-state regulation gives way to the regulation afforded by ionised gases at very low pressures. The use of neon and hydrogen discharge devices is beyond the scope of this article, but basic details may be found in reference [3].

REFERENCES

[1] *ARRL Handbook*, 1998 edn, p11.15.
[2] *Radio Communication Handbook*, 6th edn, RSGB, 1994, p3.13.
[3] *The Services' Textbook of Radio*, Vol 3, HMSO, 1963, pp180–198.

A BI-DIRECTIONAL WATTMETER

Three members of a radio club are building 80m CW transceivers for holiday use. A power / SWR meter was needed that could operate at low power and did not need to be adjusted every time the frequency or power was changed. This is the design that was chosen.

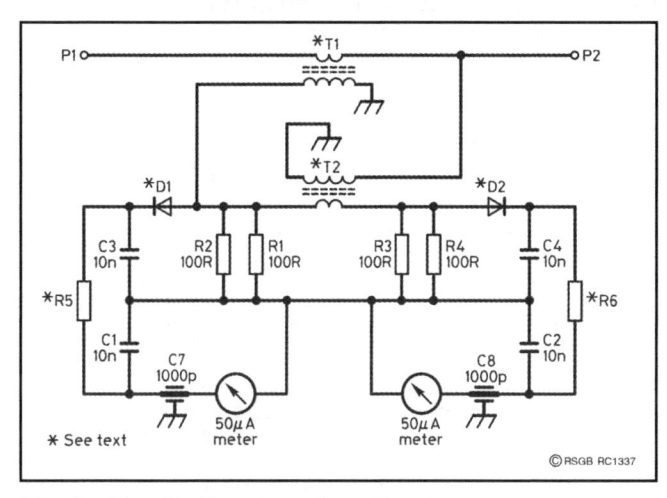

Fig 1. Circuit diagram of wattmeter.

HOW IT WORKS

When the transmitter is connected to P1 (Fig 1) and the aerial to P2, 99% of the transmitter power arrives at P2 to go to the aerial. The other 1% is sampled by the transformers and fed to R1 and R2. The RF voltage across R1 / R2 is rectified by D1 and fed to meter M1. The phasing of the windings on transformers T1 and T2 cancels any voltage on R3 and R4.

If the aerial does not present a perfect 50Ω impedance, some power will be reflected. This reflected power is 180° out of phase with the transmitted power. About 1% of this reflected power is sampled by the transformers and fed to R3 and R4. The RF voltage across R3 and R4 is rectified by D2 and fed to meter M2. The phasing of the windings on transformers T1 and T2 cancel any voltage on R1 and R2.

CONSTRUCTION

The power meter RF components must be constructed in a metal box (or a box made from PCB material), as shown in Fig 2. The transformers are wound on ferrite beads (Fair-Rite 26-43006302, Bonex 6302 PA balun bead), which are fairly large and are easy to handle.

Each transformer has 14 turns of 26SWG enamelled copper wire wound tightly on the core with the turns equally spaced. Ensure that both transformers are wound with the same number of turns. The second winding consists of a short length of thin coax threaded through the centre, one end of its braid connected to ground, and the other neatly trimmed off and left disconnected.

A BI-DIRECTIONAL WATTMETER

The braid screens the two windings as far as the electrostatic field is concerned but will not hinder the electromagnetic field. (This is known as a *Faraday cage* or *screen*).

If 'junk box' meters are used, it will be necessary to calibrate the unit using a known power meter. This is not difficult to do, provided your transmitter can be reduced progressively in output to about 1W. This is the procedure to follow.

CALIBRATION
Connect the transmitter output to your power meter, the output of which must then be connected to the reference power meter which in turn must be connected to a 50Ω dummy load. A 100kΩ variable resistor is connected in place of R5 and the transmitter adjusted to give an output of 10W, as indicated on the reference meter. The 100kΩ pot is then adjusted for full-scale deflection (FSD), its value is measured and R5 and R6 replaced by resistors of the correct values. By progressively reducing the output power a graph of calibration can be made for the meter and from that a new meter scale can be laid out.

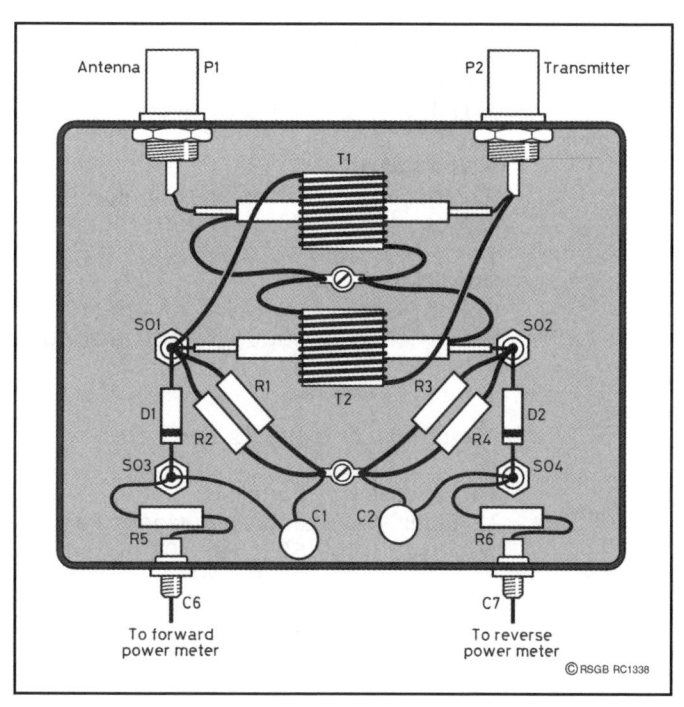

Fig 2. Component layout (meters not shown).

Making scales is easier than most people realise, provided a photocopier with a zoom facility is available. Remove the old scale and photocopy it onto paper at four times its original size. On the photocopy, the new scale is then carefully marked over the old scale using Letraset figures and lines, and all unwanted markings are covered with Tipp-Ex. Now, rather than reducing the size by four on the photocopier, reduce it by two and then touch out any remaining imperfections using Tipp-Ex, then reduce by two again to regain the original size. Using either spray-on adhesive or thin double-sided tape the new scale is stuck over the old scale and re-mounted in the meter casing. This power meter is designed for 10W, but the 43 mix beads can be used up to 100W without problems. Calibration is carried out the same way but it is necessary to increase the power rating of R1, 2, 3 and 4. Two 100Ω $^1/_8$W resistors were used to give 50Ω at $^1/_4$W for 10W. For the 100W test, four 100Ω resistors were connected in series / parallel to give 50Ω at $^1/_2$W. On SSB they do get warm, so if SSTV or very-high-duty-cycle modes are to be used, it may be advisable to increase the power rating further.

163

WEEKEND PROJECTS

PARTS LIST
Resistors
R1-4   100R carbon or metal film

Capacitors
C1-4   10n disc ceramic
C7, 8   1000p feedthrough

Additional items
T1, 2   14 turns of 26SWG copper wire on a Fair-Rite 26-43006302 ferrite ring

For other components see text.

The address of Bonex is:
12 Elder Way, Langley Business Park, Slough, Bucks SL3 6EP.
Tel: 01753 549 502; Fax: 01753 543 812.

A STANDING-WAVE INDICATOR FOR HF

The standing-wave ratio (SWR) meter shows how well the aerial system, including the feeder, is matched to the output of the transmitter. This design does not measure SWR, but it gives an indication of when the SWR is minimum for a given system of aerial and feeder. The design is usable on the HF bands from 1.8 to 28MHz, and can be used at 50MHz with reduced sensitivity.

HOW IT WORKS

There are two types of wave in any feeder: the forward wave, which travels from the transmitter to the aerial; the reflected wave, which travels back to the transmitter from the aerial. The presence of a reflected wave is evidence that some of your transmitted power is not being radiated, but is being returned to the transmitter to be lost as excess heat. When aerial and feeder are perfectly *matched*, there is no reflected wave, and all the power from the transmitter is radiated.

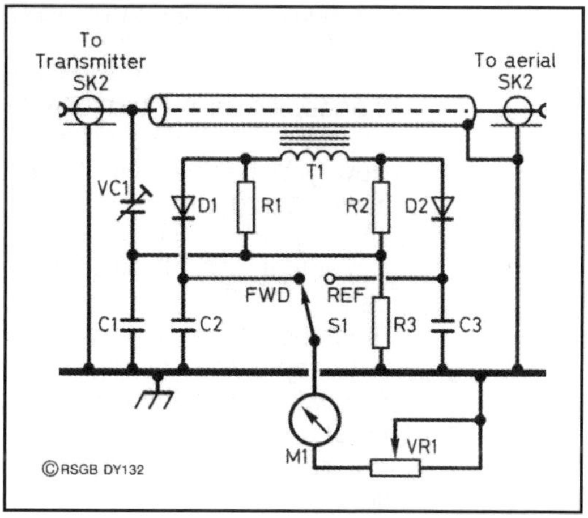

Referring to the circuit of Fig 1, a tiny fraction of the signal is removed by the transformer, T1, and by the capacitors, VC1 and C1. It is then detected by the germanium diodes, D1 and D2, and any residual RF removed by the capacitors, C2 and C3. The currents through the diode and meter (depending on the position of switch, S1) represent the forward and reflected signals. VR1 acts as a sensitivity control for the meter.

Fig 1. Circuit diagram of the SWR meter.

It pays to shop around for a suitable meter. Surplus types from tape recorders and Hi-Fi equipment are usually ideal for this purpose. A new one would cost several pounds. The more sensitive the meter, the more sensitive your indicator will be. Meter sensitivity is given by the current required to give full-scale deflection (FSD) of the pointer. One with an FSD of between 50 and 200µA is suitable for this circuit. The higher the FSD, the less sensitive the circuit.

CONSTRUCTION

The meter circuit and the sampling transformer (see Fig 2) are built and mounted on Veroboard of the copper-strip variety. It simplifies

WEEKEND PROJECTS

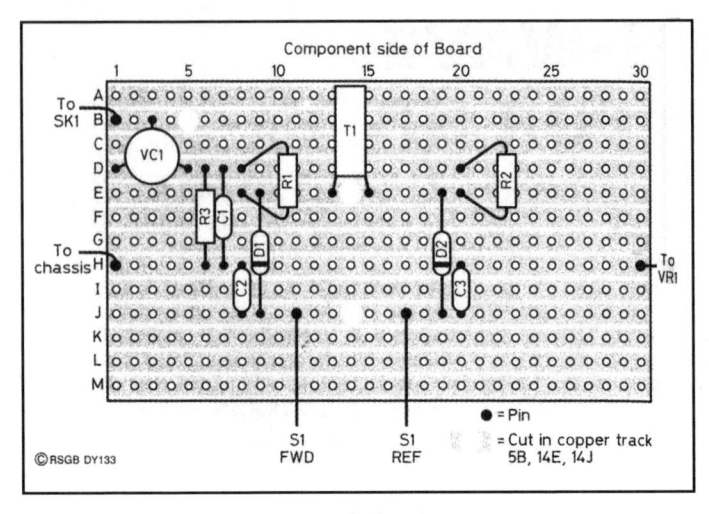

Fig 2. Component layout of the SWR meter.

construction, but reduces the operational range of the meter to below 30MHz because of the capacitive coupling between strips. The board has 13 strips by 30 holes, although you can reduce this if you have a smaller case.

Firstly, cut the tracks at the three points shown. Then insert and solder Veropins for connections to the external components, the switch, variable resistor and the meter. Solder in the components starting with the resistors and followed by the capacitors and the diodes, ensuring that the diodes are inserted correctly.

Now you have to wind the transformer, T1, on a small toroidal ferrite core. Wind the secondary with 15 turns of 36SWG enamelled copper wire, spaced evenly over about two-thirds of the former. [When winding a toroidal coil, count one turn for each time the wire goes through the centre. – *Ed*.] The turns should not overlap, and considerable care must be taken; the wire is very thin, will kink easily and will break if you apply too much tension. The primary is an 8cm length of 50Ω coaxial cable which passes through the toroid on its way between the input and output connectors. The braid of the cable is connected to the case at only one of the connectors (see Fig 1); this prevents the screen and the metal case between the two sockets forming a single, shorted turn.

The ends of the secondary winding must be carefully stripped of their enamel with sandpaper, before attaching the toroid to the board with cotton or nylon fishing line. On no account must wire be used for this!

Solder the secondary connections of T1 to the board and thread through the coaxial cable ready for soldering to the connectors.

The case used was an aluminium box (Maplin LF02C), but any suitable metal box could be used. Aluminium is preferable, as it is easily drilled with simple tools. Use standoff insulators to mount the board in the case. Once this has been done, the leads from the board to the chassis-mounted components can be soldered. So can the coaxial cable passing through the toroid. Make the lead from the input socket to VC1 as short as possible.

SETTING UP
You will need a 50Ω dummy load and a transmitter to set up your indicator. Connect the transmitter to SK1 and the dummy load to SK2.

A STANDING-WAVE INDICATOR FOR HF

Set the toggle switch, S1, to 'forward' and the sensitivity control, VR1, to mid-travel.

Switch on the transmitter, and set VR1 for maximum meter deflection. Switch to 'reflected' and adjust VC1 until the reading is minimum (ideally zero). This completes the setting up!

USING THE INDICATOR

For setting up an aerial, connect your circuit between the transmitter and your ATU, after which the coaxial cable passes to the aerial. With S1 in the 'forward' position, key the transmitter and adjust VR1 for maximum reading on the meter. Switch to 'reflected', and then adjust your ATU to give minimum reflected power. If your adjustments are to be made to the aerial itself, to give minimum reflected power, you must make a note of the reflected reading, switch off the transmitter, make a change to the aerial, key the transmitter, and note whether the reflected power is greater or less than before. Then, make more changes to the aerial. Never adjust your aerial with the transmitter on. Make your adjustments on an unused frequency, and do it as quickly as possible, thus avoiding (or minimising) interference to other stations.

PARTS LIST
Resistors: all 0.25W, carbon 5% tolerance (or Maplin 0.6W metal film)

R1, R2   27Ω
R3   2.2kΩ
VR1   10kΩ linear

Capacitors
C1   220pF disc ceramic 50 VDC
C2, C3   0.1µF disc ceramic 50 VDC
VC1 20pF trimmer

Semiconductors
D1, D2 OA91 germanium

Additional items
S1 Single-pole changeover (SPDT or SPCO)
SK1, SK2 Coaxial sockets to suit station standards
Veroboard – 13 strips by 30 holes
Veropins (7 off)
Amidon FT 50-43 ferrite toroid
Meter, less than 200µA FSD
36SWG enamelled copper wire
Short length of UR-43 or RG-58 coaxial cable
Insulated stranded wjre
Aluminium box
Standoff insulators for mounting the board
Knob for the sensitivity control

For the **best** selection of **Amateur Radio books**

Only £17.99
plus p&p

RF Design Basics
By John Fielding, ZS5JF

RF Design Basics is the latest book by acclaimed author John Fielding, ZS5JF. This book is a practical guide to Radio Frequency (RF) design rather than the more usual text book written for post-graduate electronics engineers. Aimed at those who wish to design and build their own RF equipment, this book provides a gentle introduction to the art and science of RF design.

The fourteen chapters of *RF Design Basics* cover subjects such as tuned circuits, receiver design, oscillators, frequency multipliers, design of RF filters, impedance matching, the pi tank network, making RF measurements, and both solid-state and valve RF power amplifiers. One chapter explains the meaning of S parameters, while another is devoted to understanding the dual gate Mosfet. Much attention is given to the necessity of cooling valve PAs and there is even a practical design for water cooling a large linear amplifier, a subject overlooked by most other publications.

RF Design Basics neatly fills the gap between a beginner's 'introduction to radio' and RF design text books. Written for the average radio amateur, this book is an accessible and useful reference work for everyone interested in RF design.

Size 210x297mm, 192 pages   ISBN 9781-9050-8625-2

Radio Society of Great Britain
Lambda House, Cranborne Road, Potters Bar, Herts, EN6 3JE Tel: 0870 904 7373 Fax: 0870 904 737

www.rsgbshop.org

RSGB SHOP

REFERENCE

Baluns	170
How the cathode-ray tube works	175
How the cathode-ray oscilloscope works	179
Diodes for protection	182
One-man holiday dxpedition	185
Getting started on a shoestring	189
A guide to HF contesting	193
Iota – a beginners' guide	204
Radio-frequency mixing explained	208
Noise-reduction circuits	210
The photometer and the polar diagram	213
The QSL bureau sub-manager's tale	216
Radiation resistance	221
A beginners' guide to RTTY contests	225
Safety, operating practice and the law	229
Screening – what is it and why is it important?	234
Speech processing	238
Your first use of a repeater	242

BALUNS

> The prime purpose of a balun (a contraction of '**bal**anced-to-**un**balanced') is to allow an unbalanced source to drive a balanced load or vice versa. Some types of balun will also yield an impedance transformation but this should be regarded as a secondary function.

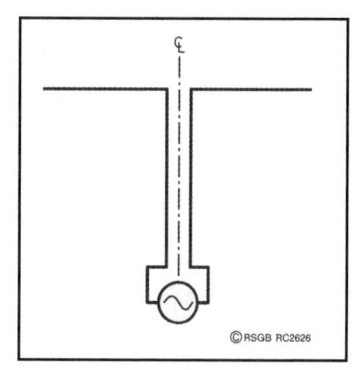

Fig 1. The standard dipole is electrically symmetrical about the centre line.

BALANCED SYSTEM

Before getting into the details of baluns, it is necessary to understand just what is meant by a balanced load, and why feeding such a load from an unbalanced source can create problems.

Fig 1 shows a typical balanced load. The arrangement is symmetrical about the centre line. Each point on the left-hand side is mirrored by an equivalent point on the right-hand side, where the currents and voltages are equal in amplitude but opposite in phase.

In the dipole itself, the currents in the two legs create fields which add together to generate the usual 'figure of eight' radiation pattern. The fields generated by each half of the feeder, though, cancel out each other so that there is no radiation from it.

FEEDING VIA COAX

Now consider the same dipole fed through a length of coaxial cable from a typical transmitter, as shown in Fig 2(a).

The current flowing in the inner conductor of the cable has only one destination, the left-hand leg of the dipole. That flowing in the outer of the cable, however, has two destinations – the right-hand leg of the dipole and back down the outside of the cable to ground.

Fig 2(b) shows a somewhat simplified representation of the various current paths, with $I3$ being that flowing back down the outside of the coaxial cable. The result of having a path for $I3$ is that a top-fed vertical aerial is, in effect, put in parallel with the right-hand leg of the dipole. This vertical aerial, will of course, radiate.

The amplitude of $I3$ is dependent upon the length of cable being used. If it is an odd multiple of a quarter-wave, the feed impedance of the effective vertical aerial is very high so $I3$ is low. Under these conditions the unbalancing effect on the dipole is insignificant. If, though, the cable is a multiple of a half-wave, the feed impedance is low and $I3$ is therefore high. As would be expected, the unbalancing effect is also high.

POTENTIAL PROBLEMS

Having explained the situation, the obvious question to be answered is: does it really matter? In order to answer this question, we need to consider two separate aspects: first, the effect on the aerial's ability to radiate; second, what might be called the *system effects* – such matters as EMC, VSWR measurement and so on.

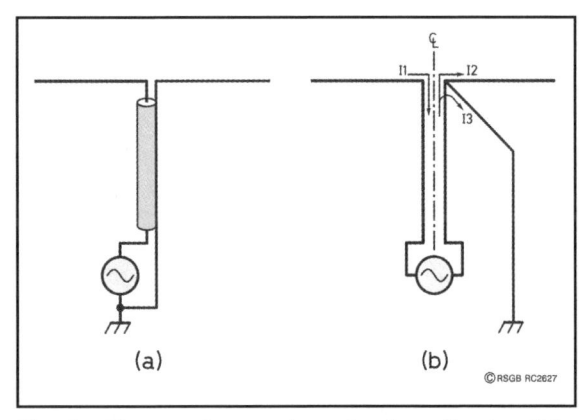

Fig 2. If coax is used to feed a dipole, this generates a current path to earth, effectively resulting in a top-fed vertical antenna.

Feeding a normal wire dipole via coax causes some change to the radiation pattern but this may not be particularly significant. There can in fact be some benefit, in that radiation from the vertical aerial represented by the outer of the coax fills in to some extent the nulls off the ends of the dipole. Should the dipole be part of an array, a Yagi beam for instance, the effects of feeding directly via coax can be quite significant. The forward gain will be reduced, as will the front-to-back ratio, and some side lobes will also appear. Similar problems will also occur with other forms of array based on balanced radiators such as the cubical quad.

Of the system effects, EMC is probably the one of greatest concern. In normal installations the coax will enter the shack and be in close proximity to house wiring. Radio-frequency energy radiated from the outer of the coax is highly likely to find its way into all the surrounding wiring and cause breakthrough problems.

Under normal circumstances, the VSWR on a transmission line is dependent only upon the load impedance and the characteristic impedance of the line. Changing the line length should not cause the VSWR, as indicated by the usual VSWR meter, to change. Sometimes, though, it is found that the VSWR reading *does* change quite significantly with line length. This variation of reading is indicating that there is something odd about the setup. Feeding a balanced load via coax represents one of the possible oddities. Fig 2(b) shows that the actual load is made up of the dipole plus a top-fed vertical aerial. Also, as mentioned previously, the level of $I3$ is dependent upon the line length. In effect, then, the load impedance at the aerial terminals varies with line length. With VSWR being dependent upon this load impedance, varying the line length will cause the VSWR meter reading to change.

It is also possible that $I3$ flowing back down the outer of the coax will affect the operation of the VSWR meter, thus causing an additional source of confusion.

So, having established that feeding a balanced load via coax causes the radiation pattern of the aerial to change, the EMC threat to increase

WEEKEND PROJECTS

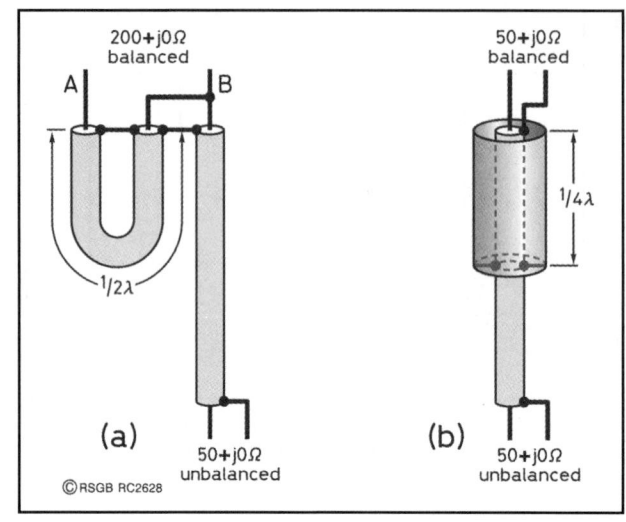

Fig 3. Narrow-band baluns can be made quite simply. A half-wave of coax, connected as in (a), generates a 180° phase shift. A quarter-wave sleeve, (b), effectively decouples the last part of the coax.

and the measurement of VSWR to become somewhat unpredictable, the next question is: can anything be done to improve the situation? Fortunately the answer is 'yes', and one of the solutions is to use a balun.

NARROW-BAND BALUNS

There are many types of balun, each having its advantages and disadvantages. Let's look first at some narrow-band arrangements – these being suitable for single-band aerials.

Fig 3(a) shows a very simple arrangement using an additional half-wavelength of coax. One feature of a half-wave line is that the voltage at the output is equal in amplitude to that at the input but with the phase reversed. With the arrangement shown, the voltages at A and B are equal in amplitude but of opposite phase. The voltage between A and B is twice that of the input. As a result, the load impedance for a 50Ω input impedance must be 200Ω. It should be noted that the aerial is not connected to the outer of the coax. If the inner of the coax does not have a DC path to earth, there can be problems with static build-up, particularly when there are electrical storms in the locality.

If a 1:1 impedance transformation is required, the arrangement shown in Fig 3(b) will suit the need. This uses a quarter-wave sleeve effectively to decouple the last section of the coax. Note that the top end of the sleeve needs to be well insulated and the bottom end connected to the outer of the coax. Although these two baluns have the virtue of simplicity, they are only really suitable for the higher frequencies owing to the lengths of line needed to make them.

BROAD-BAND BALUNS

These can be one of two basic types: those which force the currents in the two halves of the aerial to be equal in amplitude but of opposite phase and those which force the voltages to have this relationship. If the aerial is truly balanced, both achieve the same effect. One problem with wire aerials at the lower frequencies is that it is difficult for many reasons to achieve a fully balanced arrangement. The current balun ensures that in such cases the currents in both conductors of the feeder are equal in amplitude.

Fig 4 shows two simple types of current mode balun, both of which provide a 1:1 impedance ratio. These work on the principle of providing an impedance to restrict the flow of an out-of-balance current. In the case of Fig 4(a) this is achieved by coiling up the coax near to the feed-point of the aerial. The coil has no effect on the normal signal flowing up the coax but looks like an inductance to any current trying to return via the outer. The same effect is achieved in Fig 4(b) by threading ferrite rings over the coax.

One difficulty with the arrangement shown in Fig 4(a) is that it can be difficult to make it work effectively over a wide frequency range – say, 3.5 to 30MHz. Providing sufficient turns to cope with 3.5MHz is likely to result in the inter-winding capacitance being too high for effective operation at 30MHz. The situation can be improved by winding the turns on a ferrite rod or ring.

A current-mode balun can also be constructed by using a bifilar winding on a toroid or ferrite rod, as shown in Fig 5. It must be understood that these bifilar windings act like transmission lines, which can limit the performance of the arrangements in some circumstances and yield rather erratic results. Ruthroff [1] advocated the addition of a third winding to the simple bifilar winding of Fig 5(a) to yield the arrangement shown in Fig 6(a). Although this third winding overcomes some of the problems of the two-winding arrangement, it has the effect of turning the balun into a voltage-mode device. Windings 1 – 3 and W3 act like an auto-transformer, so that the voltage at point A is half that of the input. The voltage at point B will also be half that of the input but with a phase reversal. This arrangement tends to be regarded as the 'standard' for 1:1 impedance ratio baluns.

The simplest form of voltage-mode balun, albeit with a 4:1 impedance step-up, uses the same basic arrangement as Fig 5(a), but with the windings connected in a different way, as shown in Fig 6(b). Construction is as for the examples in Fig 5(b) and Fig 5(c).

There are two (often conflicting) criteria associated with the design of voltage-mode baluns. First, the inductive reactance of the windings should be high; second, the leakage reactance should be low compared with the load impedance in each case.

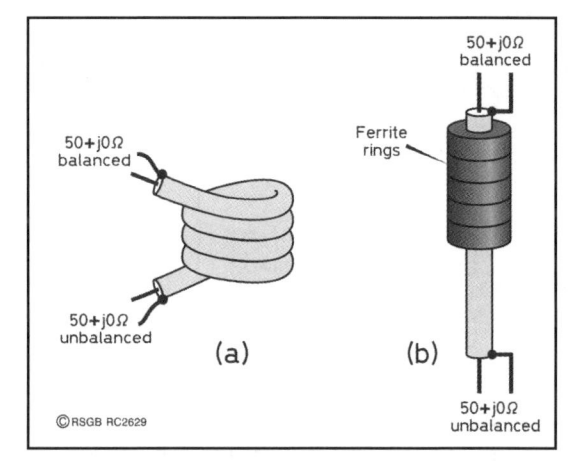

Fig 4. Two simple current-mode baluns. Both use inductance to reduce the current on the coax outer.

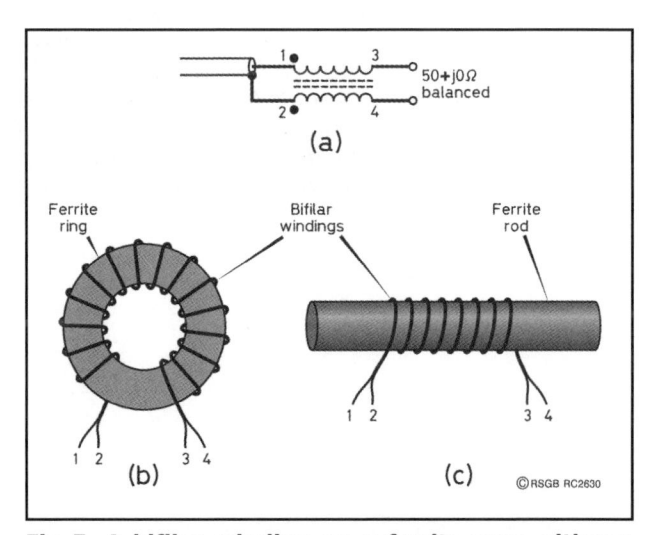

Fig 5. A bifilar winding on a ferrite core, either a toroid or a rod, yields a simple current-mode balun. Note that the dots indicate the starting points of each winding.

WEEKEND PROJECTS

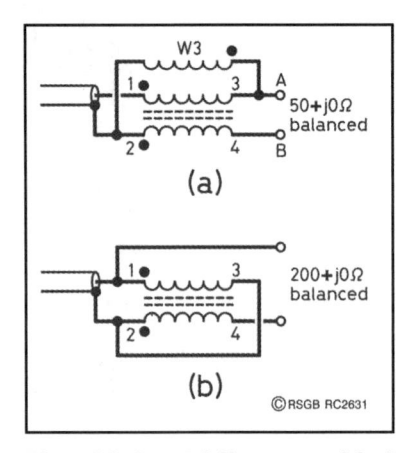

Fig 6(a). Adding a third winding, W3, to the simple current-mode balun yields a 1:1 voltage-mode balun. (b) Connecting the two bifilar windings in a different way gives a 4:1 impedance step-up design.

The first usually determines the low-frequency limit of operation whilst the second determines the high-frequency limit.

CONCLUSIONS

This has been a fairly brief introduction into the subject and has (quite deliberately) begged the question of which design is 'best'.

The reason for this omission is that the use of baluns tends to result in compromises having to be made – what works well in one application may be a total failure in others. The sensible approach is to try a few different ideas and select that which gives the best performance in your particular set-up. Fortunately, the components used are reasonably inexpensive and can easily be recycled for the different arrangements.

For those wanting to have a go, references [2 – 7] list articles and books containing more details of the different arrangements. You might be bemused by the fact that some authors will be enthusiastic about a particular arrangement whilst others regard it with horror. Take due note of any objections to the various designs but do not let these put you off trying them.

REFERENCES

[1] 'Some broad-band transformers', C L Ruthroff, *Proc IRE*, Vol 47, August 1959.
[2] 'Balance to unbalance transformers', Ian White, G3SEK, *Radio Communication* December 1989 (highly recommended reading).
[3] *Radio Communication Handbook*, RSGB.
[4] *ARRL Handbook*, ARRL.
[5] *HF Aerials for All Locations*, Les Moxon, G6XN, RSGB.
[6] *Backyard Aerials*, Peter Dodd, G3LDO, RSGB, 2000.
[7] *Transmission Line Transformers*, ARRL.
[8] *Reflections, Transmission Lines and Aerials*, ARRL.

HOW THE CATHODE-RAY TUBE WORKS

The cathode-ray tube (CRT) is the most flexible measuring instrument in the electronics laboratory. Its construction is relatively simple, and has remained essentially constant since its invention by Braun in 1897. A selection of some different types and sizes is shown in the photograph.

DESCRIPTION
The basic oscilloscope CRT is shown in Fig 1, using its standard circuit symbol, which is very similar to its actual construction. Very much like an electronic 'ship-in-a-bottle', all of its electrodes are contained inside an evacuated glass envelope.

Electron emission
The *cathode* emits a copious supply of electrons when raised to red heat by the *heater*. Only a minute fraction of these is allowed past the *grid,* a metallic cylinder capped at one end with a plate containing a circular aperture. This electrode is maintained at a potential which is negative compared with that on the cathode, and tends to repel the electrons back towards the cathode. Those electrons which pass beyond the grid are thus controlled by the potential on the grid, which is varied by the 'brilliance' (or 'intensity') control on the front panel of the oscilloscope.

Anodes
Beyond the grid (at a distance much greater than that between the cathode and the grid) are three more circular plates, each with a small aperture in its centre. These are called *Anodes 1, 2* and *3* (in a direction away from the grid) and have very special roles to play.

Anode 1 and Anode 3 are usually connected together and are maintained at a potential of between +600V DC and +5000V DC relative to the cathode; the exact voltage depends upon the design of the equipment operating the CRT and its purpose. This extra high tension (EHT) on

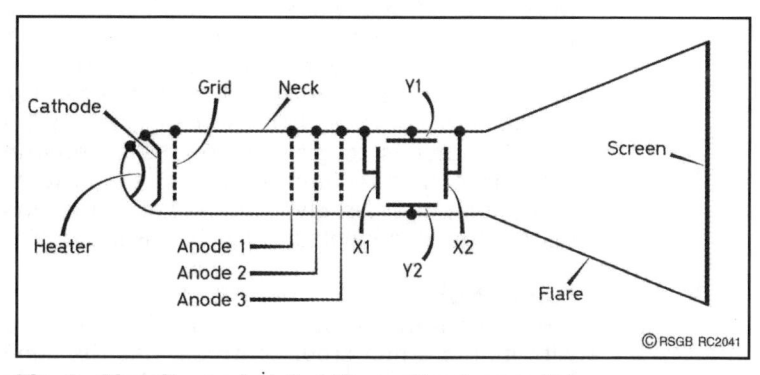

Fig 1. Circuit symbol for the cathode-ray tube.

WEEKEND PROJECTS

the anodes attracts those electrons which have passed beyond the grid and accelerates them to very high speeds (comparable with the speed of light – see later). The beam of electrons diverges slightly as it approaches the anode structure, and is travelling so fast that it passes through the holes in the three electrodes, rather than colliding with them. Anode 2 is maintained at a lower positive potential than Anodes 1 and 3 (usually around +150V DC), and its purpose is to 'bend' the electron beam so that it converges rather than diverges. The amount of convergence is controlled by the potential on Anode 2, this being varied by the 'focus' control.

The electrodes considered so far form two groups within the cathode-ray tube. The heater/cathode/grid combination is known as the *electron gun*, as its purpose is to 'fire' electrons towards the anodes. The three anodes are known collectively as the *electron lens*, and are required to focus the electron beam on to the screen. Turning the focus control changes the focal length of the lens.

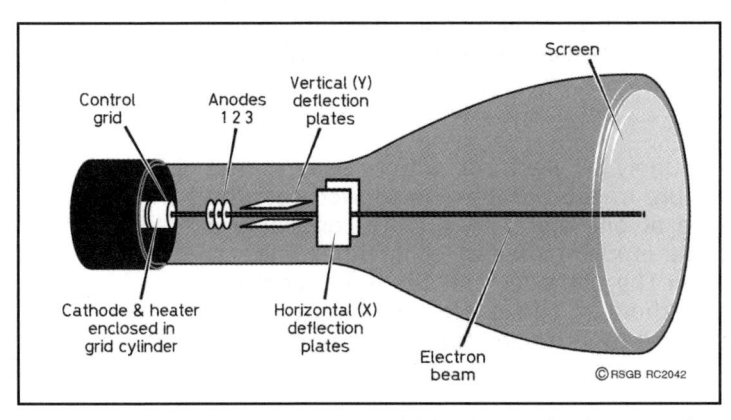

Fig 2. The electrode structure within the cathode-ray tube.

Deflection

Beyond the electron lens are two parallel pairs of *deflector plates*, as shown in Fig 2, which is *not* to scale. First come the *Y-plates*, two horizontal plates between which a potential can be established; this will exert a transverse force on the beam passing between them and thus will deflect the beam up or down (ie along the y-axis), depending upon the polarity of the voltage applied. The *X-plates* then perform an identical function except that the deflection produced is horizontal (along the x-axis).

Display

Once clear of both pairs of deflector plates, the beam continues on its way to the screen at a constant speed and direction. The screen is simply a coating of phosphor powder on the inside of the glass envelope at the end of the *flare* (which is where the tube increases in diameter between the neck and the screen), as Fig 2 shows. The purpose of the phosphor is to glow under the bombardment of electrons – the more intense the bombardment, the brighter the glow. Different phosphor materials give different colours, and the simplest phosphor material is zinc sulphide, which glows green.

Connections

The cathode-ray tube is a form of thermionic valve and, in common with its more numerous brethren, has the connections to its electrodes brought out to a series of pins in the tube base. The base pins mate

HOW THE CATHODE-RAY TUBE WORKS

with a socket, to which the connections are made from the rest of the equipment of which the CRT is a part.

If it is assumed that no electrons strike the electron lens on their way through, there is no circuit present, as there is no conventional anode to collect the electrons. To overcome this problem, and to prevent the screen charging up, the insides of the neck and flare are usually coated with colloidal graphite which, in turn, is connected to Anode 3.

Three very different types of cathode-ray tube. Left is the VCR97, an oscilloscope tube widely used in radar equipment before and during WWII, using an anode voltage of around 2kV. Centre, the Cossor 23D, another oscilloscope tube using 800V anode potential. Both of these are to be found quite frequently at radio rallies. Right is the Mullard MW6-2, a black-and-white projection television tube with magnetic focusing and deflection, using 25kV anode voltage.

This collects the secondary electrons emitted when electrons strike the screen, thus completing the circuit.

REFINEMENTS

There are two main changes to the structure which are found in more sophisticated CRTs.

First, to minimise the capacitance between the pairs of deflector plates, their connections are brought out on the neck of the tube, thus permitting the use of the plates at higher frequencies. Second, in order to increase the brightness without degrading the deflection sensitivities, another electrode *after* the deflector plates is used to provide additional acceleration of the beam. This is known as *post-deflection acceleration* (PDA).

The deflection of the beam, as measured in millimetres horizontally or vertically on the screen, is directly proportional to the voltage between the corresponding deflector plates; a doubling of deflection indicates a doubling of voltage. It is this property which

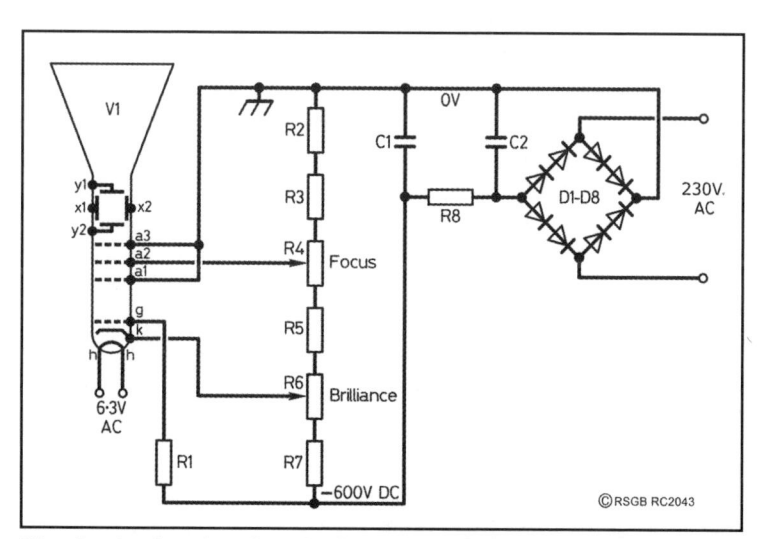

Fig 3. A simple circuit for biasing the electrodes of a cathode-ray tube.

177

WEEKEND PROJECTS

makes the CRT such a valuable tool for the measurement of very complex waveforms. Fig 3 shows how the CRT is biased.

SPEED

Finally, it is worth mentioning the speeds at which the electrons travel, even in modest systems, such as may be constructed for amateur use. For an EHT of only 600V, the electrons strike the screen at around 1/20 the speed of light; at 5000V, this becomes 1/7 the speed of light!

HOW THE CATHODE-RAY OSCILLOSCOPE WORKS

In the previous article, the principle of the cathode-ray tube (CRT) was discussed. Now it is time to look at a particular piece of equipment which uses the CRT. In the radio amateur's shack, this equipment is usually the cathode-ray oscilloscope (CRO).

BACKGROUND

In that article, the functions of the heater, cathode, grid, anodes, deflector plates and phosphor were explained, together with the operation of the brilliance (or intensity) and focus controls. It is now instructive to see how signals external and internal to the CRO are fed to the CRT, and to understand why the oscilloscope is so useful as a measurement and test tool.

X AND Y

For 90% of its everyday uses, the CRO presents a real-time display of voltage (along the y-axis) against time (along the x-axis). The varying voltage on the y-plates is usually an amplified version of an external signal applied to the y-amplifier through a socket on the front panel (see Fig 1). This amplifier has a large, fixed gain, but is preceded by a wide-range frequency-compensated attenuator in order to accommodate the display of signals of widely varying magnitudes. The calibration scale on the y-attenuator enables the signals to be measured. The signal present on the x-plates varies uniformly with time, and is generated internally by circuits known as the *timebase* and the *x-amplifier*.

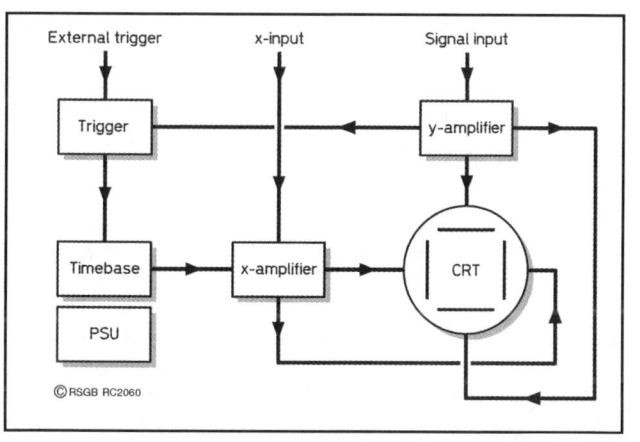

Fig 1. Block diagram of the basic cathode-ray oscilloscope.

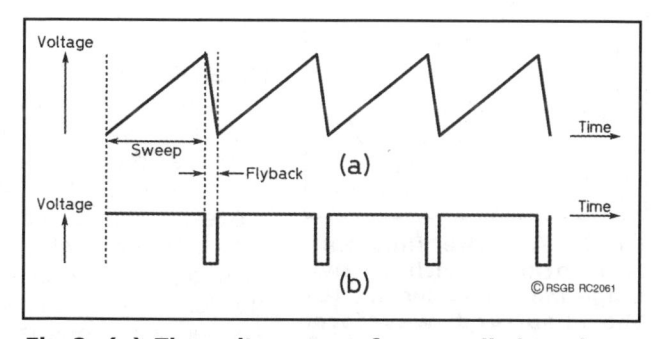

Fig 2. (a) The voltage waveform applied to the x-plates of the CRT. (b) The blanking pulses applied to the CRT grid to suppress the appearance of the flyback on the screen.

179

WEEKEND PROJECTS

THE TIMEBASE

In order to 'draw' the spot at a constant speed across the face of the tube, a *sawtooth* waveform is used, as shown in Fig 2(a). The waveform consists of two parts: the sweep is a rising voltage which is linear with respect to time and deflects the spot from left to right across the tube screen; the flyback is the falling section, which should occur very quickly, returning the spot to the left-hand side of the screen.

This process is repeated to give the continuous display of the input waveform. At very high sweep speeds, the flyback may occur in a time comparable with the sweep; when this occurs, the flyback becomes visible and clutters the display. To overcome this problem, *flyback blanking* is used. When the flyback occurs, the negative-going blanking pulse shown in Fig 2(b) is applied to the grid of the CRT; this effectively switches off the beam, and no flyback is visible. Referring to the tube bias circuit given in the previous article, the blanking pulse would be applied to the top end of R1 which is connected directly to the grid of the CRT.

The speed at which the timebase runs is governed by a calibrated rotary switch ('coarse') and an uncalibrated potentiometer ('fine'), both on the front panel. These enable the speed of the timebase to be matched to the frequency of the incoming signal, to display a convenient number of cycles on the screen. The calibration on the timebase control allows the measurement of time intervals to be made from the display, from which signal frequencies can be calculated if required.

There are three basic modes of timebase operation: *free-running*, where the timebase runs freely at the speed set by the controls; *synchronised*, where the free-running timebase is given 'kicks', derived from the incoming signal, which return it to the left side of the screen and will give a stationary display when the sweep controls are correctly set; *triggered*, where the timebase is inhibited (stopped) and can be triggered to produce a single sweep only, giving a stationary display under all circumstances. Both the synchronised and triggered modes require level and slope-detecting circuits to operate. These functions are illustrated in Fig 3.

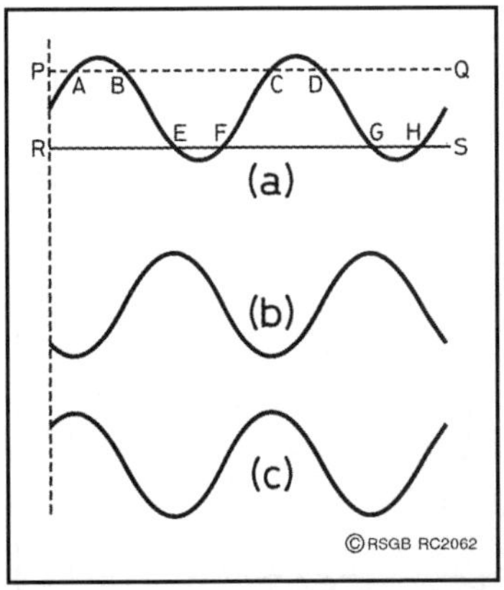

TRIGGERING

An incoming sine wave is shown in Fig 3(a). If the timebase were to be triggered (or synchronised) whenever the voltage level crossed the value indicated by the line RS, it would begin the sweep at two points in every cycle, such as E and F or G and H. If the display is to remain static (the aim of triggering), this is unsuitable. In order to isolate only one point per cycle, the detection of the waveform's slope is also needed. If a negative slope is selected, only points E and G will be found at the level RS, ie one point per cycle. The timebase will begin its sweep at the same point in each

Fig 3. (a) The incoming waveform with two triggering levels shown. (b) The displayed waveform when triggered at points E and G. (c) The displayed waveform when triggered at points A and C.

HOW THE CATHODE-RAY OSCILLOSCOPE WORKS

cycle, and thus will produce a static display, shown in Fig 3(b).

A positive slope selection would trigger the timebase at F and H. Fig 3(c) illustrates the display that would be obtained by using a positive slope at level PQ. As Fig 1 shows, it is usual to derive the triggering signal from the incoming signal via the y-amplifier, but most oscilloscopes have a socket for feeding in a trigger signal from an external source. This is most useful when viewing a complex waveform.

There is a simple test to check if a timebase is being synchronised or triggered: if the timebase stops when the signal is removed, it is being triggered. Many types of CRO do not distinguish between synchronising and triggering on the front-panel controls.

Finally, what about the 10% of cases where the x-axis is not time but is another signal generated externally? One such application is illustrated in Fig 4.

It uses the y-against-x capability of the CRO to plot the characteristics of a single component, in this case a Zener diode.

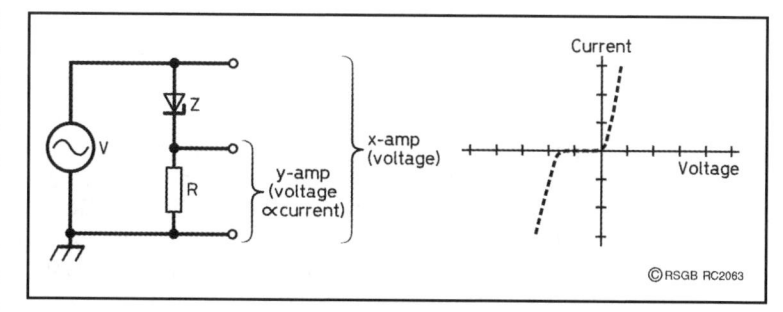

Fig 4. Displaying Zener diode characteristics using a CRO.

The voltage signal is fed to the x-amplifier, and generates the x-axis; the voltage developed across R is proportional to the current flowing, and generates the y-axis. In this case, the x-axis scale is 5V per division; the display shows a 10V reverse Zener breakdown, together with the standard forward characteristic of a silicon diode. The voltage source can be derived from a transformer or from a function generator, preferably using a triangular wave. The measurements are not source-waveform-dependent.

Other non-timebase applications include the measurement of frequency using Lissajous figures, but this has little direct application in amateur radio.

SPECIAL MODELS
Oscilloscopes made specifically for amateur radio are usually called *station monitors*; typical examples being the Kenwood SM-220 and SM-230. These are conventional oscilloscopes but have direct HF / VHF access to the y-plates, which allows a radiated RF waveform to be displayed, and trapezoidal displays of HF waveforms to be produced. They also act as panoramic displays with the TS-950 transceiver.

DIODES FOR PROTECTION

Many semiconductor devices can be destroyed in an instant if their supply is reversed With the use of batteries as power sources, it is quite common for the battery to be connected 'the wrong way round', and the scope for damaging equipment is significant and very real. Diodes can protect equipment in several ways, and you may do worse than to consider one of these approaches to protect your next expensive project.

CHOOSE WISELY!

All the diode circuits given here are so simple as to invite calamity. The circuits are not foolproof but, with a little care, will work first time. Remember that a semiconductor diode has a forward voltage drop of between 0.5V and 0.7V, depending on its type and the current flowing. This will be mentioned later.

The first thing you need to do is to insert a good multimeter in series with the circuit you want to protect; set the range to Amps DC, and switch on. Check that the circuit works properly. Then, decrease the current range on the meter until a good reading is indicated. Make a note of this current, as it is the normal running current of your circuit.

To choose a diode, you must consult the catalogues or data sheets and find one where the quoted maximum forward current exceeds the current you have measured; preferably it should be at least twice your measured current. Secondly, the diode will have a peak inverse voltage (PIV); this is the maximum voltage it can withstand when the cathode is made positive with respect to the anode, ie when it is reverse-biased and not conducting. This voltage must be greater than your battery voltage, again by a factor of about two. Except in the case of the bridge rectifier (see later), these criteria will enable the selection of a suitable diode to be made easily.

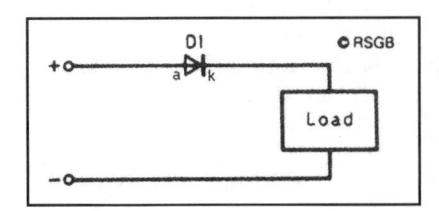

Fig 1. Series diode protection.

THE SERIES DIODE

The simplest and most obvious way to protect equipment is to insert a diode in the positive supply lead, as shown in Fig 1, with the diode passing current only when the supply is of the correct polarity Because of the 0.6V that exists across the diode, your equipment will normally operate on a slightly lower voltage. If you imagine reversing the supply to this circuit, you will see that the negative terminal of the battery is

DIODES FOR PROTECTION

connected to the anode of the diode; the diode becomes reverse-biased and will not conduct any current, thus protecting the load. Your equipment will not operate when the battery connections are reversed.

THE PARALLEL DIODE
The circuit of Fig 2 overcomes this voltage drop. It places the diode in parallel with the load (your equipment) but in a normally reverse-biased condition, so that it draws no current when the battery is correctly connected. Reverse the battery connections, however, and a very large current will flow through the diode, thus blowing the fuse! For this technique to work successfully, the current drawn by the diode when the battery connections are reversed must be much greater than the maximum current drawn by the equipment, in order to blow the fuse. This is usually not a problem, however.

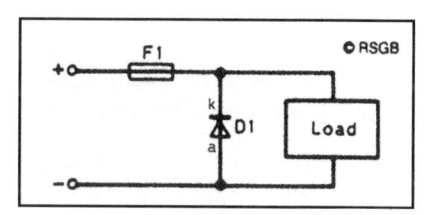

Fig 2. Parallel diode and fuse.

THE DIODE BRIDGE
For sheer elegance, the circuit of Fig 3 takes the biscuit! It uses four diodes connected in the form known as a bridge rectifier. Such rectifiers exist, and do not have to be made up from four discrete diodes.

Follow the current round the circuit from the supply, assuming initially that the top wire is positive and the bottom wire negative. It flows from the positive supply:

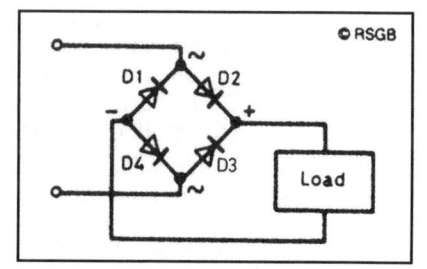

Fig 3. The diode bridge.

(a) through D2;
(b) through the load (top to bottom);
(c) through D4, and back to the negative side of the supply.

Now assume that the bottom supply lead is positive and the top lead negative. The current flows from the positive supply:

(a) through D3;
(b) through the load (top to bottom);
(c) through D1, and back to the negative of the supply.

So, whichever way round the battery is connected, the current will always flow the same way through your equipment.

The circuit does have a drawback, however. Whichever way round the battery is connected, there are always two diodes conducting the current at any instant. In the first case it is D2 and D4; in the second case, D1 and D3. This means that there is a total voltage drop of about 1.2V. If your equipment can tolerate that reduction in voltage, then you will not have a problem.

DECOUPLING
Whenever the supply rail to a piece of equipment, or even to an individual stage of a circuit, is broken for the insertion of a device that

183

WEEKEND PROJECTS

will drop voltage, strange things can happen. This is because the supply for any circuit is assumed to have a low resistance to DC and a low impedance to AC. (Impedance is the AC equivalent of DC resistance.) These two are not the same, and the insertion of a diode or diodes is certain to make a big difference to them both. To overcome any irregularities in the operation of the circuit you are protecting in any of the ways described previously, the protected supply needs to be *decoupled* with a capacitor.

If the circuit handles audio frequencies only, placing an electrolytic capacitor directly across the load in all three circuits should solve the problem. The parallel circuit of Fig 2 is less at risk than the other circuits. The size of the capacitor will be determined by the current taken by the circuit, and may need to be chosen within the range 100µF to 10,000µF, with a working voltage greater than the supply voltage.

If the circuit is mainly handling RF currents, placing a capacitor of 0.01µF across the load should prevent any problems. A second capacitor, also across the load, of between 10µF and 100µF may be needed. Again, the parallel circuit of Fig 2 is less at risk than the other two.

Don't be afraid to experiment, but confine your experimenting (at first) to small equipment and low currents, until you get a 'feel' for the technique.

ONE-MAN HOLIDAY DXPEDITION

G3LDO (ex-VQ4HX, VQ1HX, VQ3HX, 9L1HX, PA9APV, ZK1XE and CN2PD) brings us the benefit of his experience in operating 'one-man' DXpeditions from holiday venues.

In the depths of winter you (or your partner) may have booked a holiday to that seductively sunny location selected from a brochure from one of the local travel agents. If, like me, you get bored with beaches or irked by excursions after a couple of days, you could plan some amateur radio into the vacation. In this regard I am thinking more about a mini-DXpedition using HF and I am assuming that this is a 'normal' holiday with amateur radio being just a part of it. It follows that you would need a set-up that will give the maximum number of QSOs in the limited time you might have for radio operation.

Although this article is a general description of the considerations regarding taking your amateur radio equipment with you on holiday it also describes, as an example, a recent holiday to Marrakech in Morocco. Even if you don't plan to take equipment, always take a copy of your amateur radio licence. While on a world trip a few years ago I had the opportunity to operate from the QTH of a local amateur in Rarotonga, Cook Islands, and obtained the call ZK1XE on the basis of a good photocopy of my G licence.

EQUIPMENT CONSIDERATIONS
The equipment and the antenna should be reasonably efficient yet relatively lightweight. If you consider that the maximum weight allowance for a package holiday flight is around 20kg, the weight of the radio equipment is an important consideration. It should be able to operate on as many of the HF bands as possible to take advantage of the varying propagation conditions that may be encountered in a short stay.

The Antenna and ATU
Starting at the antenna, the multiband doublet would seem the most flexible and practical solution. The length is not critical (a major advantage when planning an antenna for a location that you have never seen), and a 20m centre-fed length has worked well for me.

The antenna needs to be supported above the hotel or apartment roof, preferably at the centre, where the radiation is the greatest. The lightest and most portable of masts available is the telescopic fibreglass pole, originally made for fishermen and called a roach pole. This is only 1.2m long when telescoped and can be carried in hand luggage as a

WEEKEND PROJECTS

walking stick on the vacation flight. [A suitable source is the SOTA Pole from SOTA Beams – [**www.sotabeams.co.uk/SOTAPole.htm**]].

The antenna can be installed by attaching the centre to the top of the support pole, which can be fed with 300Ω ladder-line taped to the pole. This makes an efficient antenna that is both lightweight and inconspicuous, as can (nearly) be seen in the photo. Ladder-line is good for running through gaps in doors or windows. The support pole can then be fixed to almost any convenient upright building structure using tape, tie wraps or string. An essential item when using such an antenna is a suitable ATU with provision for feeding a balanced antenna. I tried to make a small ATU that would handle 100W but found obtaining the right sized components rather expensive and difficult to source. I found a suitable moderately priced unit, the MFJ-901B, which is not much larger than an FT-817 transceiver and weighs only 600g.

The Transceiver and PSU
There are a lot of small transceivers around these days that are suitable for this sort of operation. I use an old IC-706 MkI, which has all the HF bands, plus 6 and 2m. It has a built-in CW filter and also integral SWR / power meter, which means that you don't have to take an additional instrument for tuning up the ATU. This transceiver has an output power of 100W and weighs 2.5kg - not the lightest transceiver available these days but very compact and rugged.

In the past the heaviest item needed for 100W operation was the power supply, with the transformer contributing most of the weight. This problem has been overcome with the introduction of the switch mode power supply unit (SMPSU). These PSUs have had a bad press due to the interference they can cause, particularly when installed in some television sets. However, I have to report that, when using the Watson W-25SM SMPSU, not a trace of interference could be detected, which indicates that the unit has very good built-in screening and filtering. It weighs only 1.6kg compared with a traditional transformer type, which can weigh as much as 10kg.

MISCELLANEOUS
If you are into CW, you will probably want to take your favourite paddle key. There are a couple of reasons why you shouldn't. A paddle key is usually mounted on a heavy metal base, often weighing more than 1kg, to stop it skating around the desk when you are operating. This adds unnecessarily to the overall weight. Use a lightweight paddle and fix it to a clipboard as shown in the photo. There is a further consideration. The security X-ray systems at airports are very sophisticated and can see the internal structure of radio equipment even though in a metal box. However, they have problems with thick slabs of material, such as the base of a paddle, and you may be required to identify the item. You may also wish to take a laptop computer so that you can log your contacts. However, my HP Omnibook 5700 weighs 3.5kg although some of the more modern ones may be lighter. I normally use an exercise book ruled with columns that I feel necessary; a low-tech solution that adds little to the overall equipment weight. Unless you are really *au fait* with your modern rig it is wise to take

ONE-MAN HOLIDAY DXPEDITION

some instructions of how to work through the menus and settings of your transceiver. Although I had spent some time familiarising myself with the lesser-used functions of IC-706, I found the photocopied instructions that I included did get used. The total weight of items other than the equipment described above is about 1.6kg (total weight just over 6kg).

SETTING UP

If you are lucky you will be allocated a room on the top floor of the hotel with easy access to the roof for installing an antenna. If you have been allocated a room that is unsuitable, don't be afraid to ask if you can change to a more suitable room; I have never had this request refused although I may have been lucky in as much as there were spare rooms available. Note that it is easier to make the change before unpacking suitcases! If you are staying in a hotel, establishing good personal relationships with the hotel management at an early stage is important.

The CN2PD set-up at the Hotel Tikida Garden, Marrakech. The IC-706 sits on top of the Watson W-25SM SMPSU, with the MFJ-901B ATU to the left. Note the lightweight CW paddle key fixed to the clipboard.

The chances are they won't know what amateur radio is, so some education might be necessary and to this end a copy of the licence and some amateur radio literature is useful. The main reason is the business of erecting an antenna on the roof of the hotel. During my vacation / DXpedition to Marrakech, the Reception Manager even provided a couple of maintenance men to help me put up the antenna. Do not venture on to a hotel roof without permission. Setting up an amateur radio station in an apartment is often a different matter because there may be no-one around to discuss the matter of obtaining permission. For an AC mains connection I originally considered removing the British 13A plug from the PSU and fitting a suitable continental two-pin plug (or one appropriate to the country being visited). I decided against this approach and now prefer the use of a four-way 13A extension board. Some hotel rooms have only one mains socket and the distribution connector allows the use of British electrical items such as chargers and a hairdryer, without removing the13A plugs.

Furthermore, each item is separately fused. The extension connection to the hotel mains socket can be made via a 13A plug and an adapter - this arrangement allows the whole installation to be fused. A 10A fuse is suitable on a 220 - 240VAC supply. A 13A fuse will be required for a 115VAC supply.

OPERATION

I am not that fond of pile-up and contest type operating, preferring short chats; what you might call 'rubber stamp QSOs'. However, if conditions are good to other parts of the world that regard you as DX, a pile-up will ensue. Use headphones to keep the noise in the hotel room down. I found the Heil headset with the boom microphone useful for this sort of operation. When things really got going the QSO rate could be two a minute. With my proficiency (or lack of it) on CW, the best I could do was one per minute. However, there are some real advantages for this mode for vacation operating; with headphones it allows QSOs to be made while your partner watches television. You can also sneak in an extra half-hour operation on the way back from the toilet in the middle of the night (the G3 complaint) without disturbing anyone.

THE CN2PD OPERATION
You may notice that the equipment shown in the photo includes a laptop, against the advice above. The reason for this is that it formed, together with an LF preselector and amplifier (box top left), part of some 136kHz listening experiments. The results of these experiments were described in the 'LF' column in *RadCom*, December 2003.

When it was first decided to go to Morocco, I found information regarding obtaining a CN licence difficult to obtain. The information I did receive indicated that I should write to the Agence Nationale de Réglementation des Télécommunications (ANRT) in Rabat, including a copy of my G licence and passport. I was led to understand that the fee was 300 Dirhams, about £20. In the event the correct procedure was to wait for a reply to the letter, which would include a bill for the licence fee, which is to be paid into a local bank on arrival (in this case only 70 Dirhams). Full details on obtaining a CN licence can be obtained from the RSGB Amateur Radio Department.

GETTING STARTED ON A SHOESTRING

Many people will tell you that amateur radio is an expensive hobby. It's true that it *can* be, but it is important to emphasise that it *need not* be too expensive to get on the air.

Many newly-licensed amateurs will already have a suitable receiver, having spent an 'apprenticeship' as a listener but, once their new licence arrives, they will naturally want to transmit. The logical means might seem to be to build or buy a transmitter. However, for the beginner this is not always the easiest way to get on the air. Unless the receiver has been designed to be used with a specific, separate, transmitter, it can be quite complicated to arrange the transmitter/receiver aerial changeover switching. Another difficulty is arranging receiver muting, so that the receiver is not completely overloaded by the strength of one's own transmitted signal. At worst, damage to the receiver can occur if attention is not paid to this.

A generally more satisfactory option, especially for the beginner who is impatient to get on the air, is to buy (or build) a transceiver, rather than use 'separates' (a separate transmitter and receiver).

A *new* HF, VHF or UHF transceiver from any of the 'big three' manufacturers – Kenwood, Yaesu and Icom – will cost anything from a little under £800 to £4000 or more, and they will give you every facility you need – and a lot you don't. A 'basic' transceiver, on the other hand, just needs to transmit and receive on one or more bands. Obviously there are several features that are essential – the AF gain (volume control) to use an extreme example – but many of the features found on modern transceivers, such as memory channels, speech processors and scanning – so-called 'bells and whistles' – while nice to have, are not absolutely necessary.

A Heathkit SB-101 HF transceiver is typical of 1970s equipment which is now available for around £200.

WEEKEND PROJECTS

So instead of buying a new transceiver, here are some alternative suggestions of how to get on the air cheaply. But first, you must decide which band or bands you want to operate on and this will largely be dependent on your licence class. For those looking for a VHF or UHF rig, the decision of whether to operate only FM or also to try SSB and / or CW also has to be taken into account.

SECONDHAND

A look through the 'Members' Ads' in the last couple of issues of *RadCom* shows that there are some real bargains out there. For an HF transceiver, how about a Yaesu FT-101ZD at £300 or an Icom IC-701 at £295? Still too much? A Kenwood TS-520S was listed at £225. Want to try QRP for the first time? A Kenwood TS-120V 10W PEP transceiver was listed at £275. Interested in putting on WAB (Worked All Britain) squares on 80m? A Tokyo Hy-Power HT-180 80m single-band mobile transceiver was offered at just £150. Want a multi-band aerial as well? A Ten-Tec Argosy transceiver with an HF5 five-band vertical aerial was for sale at £295.

Don't forget that these transceivers do much the same as brand-new transceivers costing 10 times the amount or more. Maybe you already have an HF transceiver but want to try some VHF DX work for the first time. Rather than buying a new dedicated VHF multi-mode transceiver, the *RadCom* 'Members' Ads' recently offered two Microwave Modules MMT 28/144 transverters, one at £75 and the other at just £50. A 10m to 4m transverter was also offered for only £50. If you're interested in 2m or 70cm FM operation, a wide variety of second-hand equipment is available: a Kenwood TR-3200 70cm portable for £130, TH-28E 2m transceiver with 70cm receive capability for £175, or a TR-2200GX 2m FM box for £110. Something for the car? A TM-201A 25W 2m mobile was for sale at £150. Want to try SSB and CW as well as FM? A Yaesu FT-290R was for sale at £215.

But what do all those numbers mean? What is the difference between a TR-2200GX and a TM-201A? Here publications such as *The Rig Guide* [1] can be of great use. This lists hundreds of rigs by manufacturer and model number, giving the modes, bands, power output and, where available, year of manufacture and original price. An even better way to find out is to have an expert to give you a bit of help. If you are a member of a radio club, almost certainly there will be a member who will know all the major manufacturers' product numbers for the past 40 years off by heart! Let him know which bands and modes you want to operate, and how much you are prepared to spend, and he should be able to let you know which rigs to look out for.

RALLIES AND JUNK SALES

The 'Members' Ads', useful as they are, are not the only way to pick up second-hand equipment. Radio rallies held around the country throughout the year frequently have a 'bring and buy' stand at which bargains can be found. Go round the rally with an experienced guide so that you don't end up buying something you later regret.

GETTING STARTED ON A SHOESTRING

Most radio clubs have an annual 'junk sale' at which real bargains are available. In many cases, prices are rock bottom, with some pieces of equipment going for a nominal sum of £1 or even less. Again, it helps to sit next to somebody who can advise you on what to look out for – and what to avoid! Just going to radio club meetings and letting it be known that you are looking out for a cheap transceiver can often bring results. Often, people have pieces of equipment which they almost never use but have never got round to advertising them for sale. Usually, they are only too willing to let it go at a good price to an enthusiastic beginner.

Former PMR equipment like this Pye Europa can often be found at junk sales or rallies.

Former PMR (private mobile radio) equipment often finds its way to club junk sales. This is equipment from firms which need to use radio to keep in contact with their employees: typically taxi firms, window cleaning companies, security firms, garden contractors and so on. PMR equipment operates on AM or FM on frequencies relatively close to the 4m, 2m and 70cm amateur bands. Some PMR equipment is quite easy to convert to amateur band use and these are often available at junk sales at very low prices. They can be both base stations and mobile units. Again, talk to your radio club expert on such matters for advice on which equipment to look out for and for help in converting them.

DEALERS

A half-way house between buying a brand-new piece of equipment and buying from a junk sale or from 'Members' Ads' is to buy used equipment from a dealer. Generally speaking, prices are somewhat higher than in 'Members' Ads', but the great advantage is that most dealers offer a guarantee, usually for three or six months, on used equipment. Most will also offer a repair service after the guarantee has run out. If you cannot afford the overall price immediately, most dealers are also licensed credit brokers and can offer a number of tempting ways to persuade you to part with your money!

A look through the dealers' advertisements in *RadCom* reveals several second-hand HF, VHF and UHF transceivers for under £300. Dealers will usually take your old equipment in part-exchange, so if you have recently gained a Class A licence and want to try HF for the first time you can trade in your old VHF gear to ease the burden of payment.

HOMEBREW AND KITS

Last, but certainly not least, one of the cheapest ways of getting on the air is still to build your own equipment. While the 'Third-Method SSB HF Transceiver' project (see *RadCom* June and July 1996) is not intended for beginners, simple transmitter circuits have been published in *RadCom*'s 'Down to Earth' column [2] and in *D-i-Y Radio* [3, 4]. The old RSGB publication *Practical Transmitters for Novices* [5], while nominally

WEEKEND PROJECTS

aimed at the Novice licensee, contains ideas suitable for anyone wishing to build simple transmitters. The book includes designs for 160, 80, 6m and microwave transmitters. But if you don't feel confident enough to build equipment from scratch, a kit may just be the answer. A number of dealers who advertise in *RadCom* sell transceiver kits, and the simple 80m transmitter described by Rev George Dobbs, G3RJV, in reference [3] is based on a kit called a 'Oner'.

CONCLUSION

By now, it should be clear that amateur radio need not be an expensive hobby. Building a rig from scratch can cost as little as a few pounds, a kit a few tens of pounds, or a second-hand transceiver with very similar performance to a new one costing a couple of thousand pounds can be yours for around the £200 mark. Make sure you get guidance from an experienced amateur before you part with your hard-earned cash, especially if you are not absolutely sure of what you are buying.

REFERENCES

[1] *The Rig Guide,* RSGB Books.
[2] '160m AM transmitter', *RadCom* April 1996.
[3] 'A breadboard 80m CW transmitter', *Electronics Cookbook*, RSGB, p182ff.
[4] 'A 6m transmitter', Ian Keyser, G3ROO, *D-i-Y Radio*, Vol 6, No 3, May-June 1996.
[5] *Practical Transmitters for Novices*, John Case, GW4HWR, RSGB. (Out of print – try your local club library)

A GUIDE TO HF CONTESTING

You don't have to be a Novice or newly-licensed amateur to want to find out about HF contests – the bug can bite at any time, even after many years of being on the air. Here you will find out how to get started for the first time and how to improve your operating skills.

COMPROMISE NECESSARY
No matter how recently your interest in amateur radio has developed, you cannot have failed to come across a contest when tuning the HF bands. There are few weekends in the year when there is no contest activity on SSB, CW, PSK31 or RTTY. Generally speaking, because of the nature of propagation, HF contests tend to be busier than VHF contests, but the same basic skills are involved and an operator who is successful in one is likely to do well in the other.

The amount of activity generated can, and does, cause controversy. Some people complain that the presence of a contest spoils their enjoyment of the hobby, as they are unable to find space to have their usual 'ragchew'. However, the very large amount of activity generated is proof of the popularity of contests and those wishing to take part feel they are entitled to the recognition of their aspect of the hobby. This is not to say that the needs of non-contesters should be disregarded but, love them or hate them, contests have been established for more than 60 years and are here to stay. The International Amateur Radio Union (IARU) decided that the WARC bands (10, 18 and 24MHz) shall be totally free of contests in order that non-contesters have completely clear bands in which to operate. Additionally, individual member societies of IARU (such as the RSGB) have established their own rules which stipulate frequency limits in order that a portion of each band is left clear of contest traffic. Probably the only remaining exceptions to this practice are the major contests, such as *CQ* World Wide, which attract truly worldwide participation and place great demand on all available space.

WHY ENTER CONTESTS?
The reasons for participation in contests are many and varied. The main reasons are to satisfy the basic competitive instinct which is inherent in the majority of us, to see how your station performs against those of your friends or rivals, the thrill of seeing your callsign in print, and in future years watching your call climbing up the results table. You will come to look forward each year to your own favourite contest, one that suits your station or interests. Whatever level you attain in contesting it will most certainly make you a more confident, proficient and skilled operator.

WEEKEND PROJECTS

Don't forget that you can *participate* in a contest without *entering* it. There will be stations on the air in a contest which you will never see at any other time. What better way to enhance your DX totals?

Not everyone can be a winner. For one reason or other it is not easy to get a level playing field – there will always be someone who has a bigger or higher aerial, greater power, better location, or better propagation. Striving to make the most of what you have is what brings the greatest satisfaction.

There are always goals and targets to be set and achieved, such as beating last year's score overall and on each individual band. In the case of RSGB contests, an equipment code has been devised which accompanies each listing in the results table. This will show you what you are up against and enable you to compare your performance against similarly-equipped stations. This code gives the height and type of aerial in use, the number of elements and the output power used. Details of this code can be found in the 'RSGB Contesting Guide' which is published in the January *RadCom* each year.

The RSGB HF Contests Committee (HFCC) has introduced restricted sections in many of our events, with the purpose of increasing the activity and competition amongst stations with the 'average' set-up, ie those who reside in built-up areas or have small plots. These restrictions relate to power output and aerial height. There are also categories for low-power (QRP) operators. Although we only produce one listing for the results, it can be quite readily seen from the results table and equipment code the category of each station, and certificates are awarded to the leading stations in each section.

HOW DO I START?

With the exception of National Field Day, there is no necessity

The Bristol Contest Group operating as G6YB/P in NFD and on 50MHz in June 1996. Aerials are 4-element beams for 21 and 14MHz, 7-element 50MHz beam, 4-element 28MHz and 2-element 7MHz beams.

A GUIDE TO HF CONTESTING

to register in advance for any HF contest that you may come across. Your participation will be most welcome and indeed desired by those you hear calling, so go right ahead and join in.

The first important thing to do is to *listen* carefully to a few contacts. Establish, if you can, what the event is and what 'exchange' is needed from you. You will need to know the basic rules and (except for the smallest of national contests) this information can be found in the current issue of *RadCom.* First, refer to the Tim Kirby, G4VXE, 'Radio Sport' column. Each month he publishes a 'Contest Calendar' for the current month, and the odds are that the contest you have heard will be mentioned there. If it is an RSGB event you can then refer to the 'RSGB Contesting Guide' (published each year in the *RadCom* January issue) for the full rules. If it is an overseas contest, reference should be made to the 'HF' column of Don Field, G3XTT, which includes brief rules of foreign contests for that month.

Most 'exchanges' require a report, eg 59 or 599. Now that computer logging has become the norm and where the 59 is already in the 'report' field, just waiting to be entered, you are not going to be too popular if you give a 46 or 478 report to a station who is working three or four contacts per minute; it is better to accept that this part of the exchange has become meaningless, but is still necessary! QRP and some other contests still require accurate reports, which is just as it should be. In addition you will, in most cases, need to include a progressive serial number, commencing at 001. If it is an RSGB contest, you may also be required to give your 'county code', which again can be found in the 'RSGB Contesting Guide'.

Instead of a serial number, you may be required to give your power, your *CQ* or ITU zone, your IOTA reference, or even your age! With only a little experience you will soon get the hang of it and if you get it wrong – no problem – the other station will tell you what is required of you.

Most operators dip a toe in the water by tuning through the band answering CQ calls. There is an undeniable thrill in raising distant stations or getting through against a number of other callers. However, unless the event is a big one, you are likely to spend an increasing amount of time searching for new stations to work, and the time then comes when you decide to get really involved by finding a clear frequency and make your first 'CQ contest' calls.

Unless you have a very-well-equipped station, don't expect miracles! However, there will come a time when you, as a British station, will be in demand and two or three callers, perhaps more, will come back at the same time. Now you're on the hook – your first mini-pile-up. Don't panic!

CONSIDERATION AND COMPETENCE
On the HF bands you may have heard an operator working from a rare country. Everyone wants to work him but very few are making it, so the frustrations of a few boil over into anger and we get the bad

behaviour of those operators who make it impossible for anybody to get a contact. A good operator keeps his temper. Competent handling keeps waiting stations in anticipation, in the belief that they have a chance of a contact. Consequently, they tend not to misbehave. The ability to handle a pile-up seldom comes as a natural ability but is gained by practice and experience. There is no substitute for listening. The next time you come across a pile-up, note how the operator deals with it, analyse your reaction to the style; are you spell-bound in admiration or are you critical?

Back to your contest now. Resolve not to be like that DX operator, listen carefully and try to pull something useful out of the pile-up. Often one caller is slightly behind the rest or just one word will often do. Saying something like "the station with Juliet in the call, you're 59 040" will isolate him, get you the contact, and, followed by a simple "QRZ?" bring in the other stations who will have waited. Uncertain handling at the outset, asking for repeats, would have sent them on their way up the band looking for more efficient stations to work.

CW CONTESTS
CW contesting is perhaps more daunting for the beginner, mainly due to the sending speed. The keyers employed in the logging programs are usually set somewhere between 28 and 32WPM. This can be a real demotivator, but do not be deterred.

For the Foundation licensee, or those newly on the air with an HF licence, the RSGB has designed a special series of contests for operators who would like to try out contesting in a series of relaxed, slow-speed events. Known as the 'Slow-Speed Cumulatives', these events are held in the spring and the autumn and comprise five sessions in each event. The duration of each occasion is 90 minutes and there is a maximum allowable speed of 12WPM.

The sessions are arranged over a five-week period and staggered one day each week, commencing on Monday the first week, Tuesday the next week, and so on. This way each day Monday to Friday is covered and, as this is likely to cover the night of your local club meeting, we should like to encourage participation from club stations, where guidance from experienced operators will also be on hand. You can enter from the club either as a single operator or use more than one operator in the 'multi-op' category. If you don't have your own station or club facilities, consider asking another amateur if you may use his station. Read the rules in the 'Contesting Guide' carefully and note the relevant power limits stated.

Having mastered the basic format, you will be ready for greater things. Speed comes with practice, so just remember that all the top-flight operators that you will come across started exactly where you are now. The other operator does not want to know your name or QTH, nor is this the time to ask for QSL information. He isn't being impolite, but he is in a contest and so time is of the essence.

A GUIDE TO HF CONTESTING

If you choose to try a few CQ calls, on no account send faster than you can comfortably receive. This may sound obvious but can easily happen if using keyboard or memory keyer sending, and the machine gun that will surely hit you can spread an awful lot of egg over your face. The last thing you need at this stage is damage to your confidence. When sending at your own speed, you will find that calling stations will more or less match your speed. Don't be afraid to ask them to "QRS PSE" if they are too fast; it's your frequency and you call the shots.

RSGB CONTESTS

Contests can be divided broadly into international and national events. International contests such as *CQ* World Wide are usually 48-hour events where everyone works everyone. The RSGB takes a different approach. Being of the opinion that there already sufficient all-band contests, it has gone the route of promoting single- or dual-band contests, with activity restricted to 12 or 18 hours' duration. There are, of course, exceptions. First, the Commonwealth Contest is on all bands, 80 – 10m, and lasts for 24 hours. As participation by definition is very limited, contact rates are very slow and a lot of skill is required to obtain a good result, and little disruption is caused to non-participants. The other exception is the RSGB IOTA Contest. This is a 24-hour all-band worldwide affair; it does not specifically promote UK stations but puts the emphasis on everyone contacting islands of the world on an equal basis. Different in format and concept from other contests this event has been warmly received in all areas of the world and is already the RSGB 'flagship' event with an entry rate growing very rapidly.

On the home front, the RSGB HF Contests Committee has been keen to foster the spirit and activity of contesting among the membership and to try to offer something to all of the different interest groups. Take a careful look at the 'RSGB Contesting Guide' and note the preamble to each of the events described there. The slow-speed contests have already been mentioned, but there is a special slow-speed segment of the band in the Affiliated Societies Contest, and the LF Cumulatives have been designed as training periods for stations wishing to progress to higher profile contesting. It is in these events that many of us started our involvement with contesting.

For SSB operators, the Club Calls Contest is primarily for club stations, although individual entrants are also welcomed, and we would like to see more clubs taking part in the 21/28MHz SSB Contest. In fact it will be noted that multi-operator categories have been added to most of our contests to encourage participation from clubs, the newly-licensed, and others who may not have their own equipment.

The contests mentioned in the previous few paragraphs are all mainly low-key events, designed to get you started. Many of you reading this now are new to the hobby and are our future, our lifeblood. All the successful operators that you hear on the bands began just like you, wondering what it was all about. The UK may not rank among the top nations in the world for quantity of contest operators, but we are right up there when it comes to *quality* and we wish to continue that

WEEKEND PROJECTS

tradition. We would like to increase the *quantity* too, for it is a fact that many countries with far smaller populations, such as Finland and the former Yugoslavian states, leave us far behind when it comes to numbers participating in the major contests. If your club takes part in NFD or SSB Field Day, *do* make an effort to be part of it. You will be welcome, and learning first hand from experienced operators is by far the best way to learn the ropes. If the club does not currently take part, why not organise something yourself from amongst the membership? The odds are that there will be at least some aerial equipment belonging to the club tucked away somewhere. You may not finish top of the table, but you will learn a lot, you will have a lot of fun and memories that will last a lifetime, and that is as important as winning.

The purpose of the RSGB HF Contests Committee (HFCC) is not only to decide policy and arrange the contest programme on behalf of the Society, but also to meet the needs and desires of the membership. Your views and feedback on what the Committee does are welcome; you are invited to contact members of the Committee for advice or guidance.

INTERPRETING CONTEST REPORTS

We try to publish useful hints and tips in the contest reports where space allows, and there are always detailed reports of NFD, SSB Field Day and the IOTA Contest. Then there is the *RadCom* 'Radio Sport' column, already mentioned, which contains current happenings, trends and developments, as well as suggestions on how to improve your station.

When reading contest reports, such as those in the 'Radio Sport' column, frequent reference will be found to tactics adopted by the operators. Some of the terms used may require explanation:

Run frequency

Finding a clear frequency, calling CQ and working a stream of callers is known as *running* and the frequency used is the *run frequency*. Unless you have a commanding signal, it will soon be found that success with this method is short-lived, for someone with a better signal will park right on top of you and he may not hear you (or, worse still, he may not wish to hear you). Whatever the reason, if he doesn't move after the first time of asking, you may as well forget it and find another spot. Good operators will move but it may be that he genuinely does not hear you. Time spent arguing the toss is completely wasted, and likely to put you in a bad mood, serving only to delay your progress in the contest. Accept the fact that this is going to happen (it happens to the 'big guns' at times so it's bound to happen to you), so the sooner you move away and start afresh the better.

When you have a good aerial or enjoy the peak of propagation on a band that is open to all areas, a good 'run' can be an exhilarating experience, limited only by the speed at which you can work callers. It is common for DXpedition-type contest stations to work in excess of 200 stations per hour.

Searching and pouncing (S&P)

This involves combing the band and checking each station heard either for a new contact or for a new multiplier. This tactic needs to be methodical and accomplished as quickly as possible. It is very easy to spend too much time calling a needed multiplier to the detriment of your overall performance, so it is necessary to have a plan of campaign.

Keep a note of stations that you don't raise after a few calls. If your rig has memories it is an advantage to store them, then you can return again at will and be bang on the frequency. So how long should you call? You need to make a judgment based upon the rules of the contest. For example, let us take two contests run by the American magazine *CQ*, probably the two busiest contests of the year.

The first is *CQ* World Wide. Suppose the desired station is SU1ER in Egypt. First, you can be pretty sure that you are unlikely to run across another Egyptian station, and it is a much-needed DXCC multiplier. In addition, SU is in *CQ* Zone 34, and in this contest zones *also* count for multipliers, so he is worth two multipliers. Experience will soon tell you that active stations from Zone 34 are also very thin on the ground.

If, however, the contest is *CQ* WPX, he would not be so important. In that contest, only the prefix counts as a multiplier – and then only once and not on each band. So a simple DJ5 prefix is equally important if you don't have one, and your time would be more productively spent looking for other new prefixes which are easier to work.

'Mults'

The importance of multipliers cannot be over-emphasised. Obtaining the maximum possible number is essential to obtain a good score. Often it is not the station with the greatest number of contacts who is the winner but the station with the most multipliers.

An illustration will make this clear: station A has 300 QSOs and 30 multipliers, station B has 250 QSOs and 37 multipliers. Awarding three points per QSO we find that A has a final score of 300 × 3 × 30 = 27,000 points, whereas B has 250 × 3 × 37 = 27,750 points and claims the prize. It is therefore a matter of striking the right balance between running and S&P. When listening to experienced operators you will notice that they don't miss an opportunity to make new multipliers and you may have heard them referring to 'moving people around'. When a rare multiplier is running a frequency, some of the callers will be heard to ask "can you QSY to 15m?". If the caller himself is rare DX there may be a positive response and you are left with an empty frequency, cursing a missed opportunity and wondering if you will find him again on that band. Your reaction should be to follow the QSY – after all, you have noted the new frequency and that is hot information. You can make it before he has time to build a new pile-up and even before the Packet Cluster followers get to him.

If the DX station calls *you*, then you must think on your feet and grasp the opportunity. Assess which other bands are likely to have a path at that time and ask him to QSY to the greatest probability band first, followed by other needed bands. If you can do this with all needed multipliers for any country or area with relatively low amateur radio activity, it will pay dividends because you may never run across them again in the contest. Desired stations can be very hard to locate on a busy band.

'Skeds'
Scheduled contacts ('skeds') made during the contest, are common practice. However, skeds made *prior* to the event would be a breach of the spirit (and sometimes the rules) of the contest. If you contact a VK on, say, 20m during the daytime, it may not be possible to move him to any other band instantly. However, you do know that there is a good probability of making a contact with that area at 1730UTC on 7MHz, so you make a sked with him and even though the going may be tough at the time of the sked, the fact that you are listening for each other can result in a QSO and multiplier that may not otherwise have been obtained.

Skeds should also be tried with other hard-to-find multiplier stations for the higher or lower bands as the case may be. Modern computer logging programs will display a warning message on the screen one minute before the sked time.

AFTER THE CONTEST
If your approach to the contest was casual, you can simply close the log on it after the event – you are under no obligation to submit an entry. However, if you made more than a dozen or so contacts, the adjudicators would appreciate a check log from you, because the more information they are given, the more accurately they can score the logs submitted.

But, if you have decided that you *do* wish to submit an entry, what do you do next? First, it will be necessary to compile the log to be submitted. Here the 'RSGB Contesting Guide' gives comprehensive details of the format required. Assuming that your first efforts are on paper, printed examples of log sheets can be found in recent editions of the *RSGB Yearbook*.

Scoring the log
Reference to the contest rules will give you the basis upon which to score your log. Each contact is allocated a basic number of points. Imagine that you have had 50 contacts spread equally over five bands with a basic points-per-contact of 5. First of all, check the contacts for each band. If you have contacted a station more than once on the same band, points should not be claimed for the subsequent contact and it should be marked "Dupe", meaning duplicate. The penalty for failure to observe this rule can be severe; RSGB policy is to deduct 10 times the number of points claimed for that contact and in this example that penalty would be equal to the entire number of points claimed for that band! The next stage is to check the multipliers or bonuses.

Mark each different multiplier. For instance, if your 10 contacts on one band include four from USA, two from Canada plus four other countries, you will have a total of six multipliers for that band. Total the points for each band, then add them all together, then do the same for the multipliers. In our example you will have 50 points for each band × 5 = 250 points for total points score. If your multiplier total is 40, you would have a final score of 10,000 points, ie 250 × 40.

If bonuses, rather than multipliers, were used, the bonus contacts would be totalled and multiplied by their value. So if the bonus was worth 10 points per contact the final score would be 250 + (40 × 10) or 650 points.

SUBMITTING THE ENTRY

All that remains to be done is to transfer this information to a summary sheet, an example of which can also be found in the *RSGB Yearbook*. This sheet, together with the log sheets, a dupe sheet, plus a list of the multipliers being claimed, can now be posted off to the organisers. Don't send it by recorded or registered mail. If you require confirmation of safe receipt enclose an SAE or postcard with a stamp or International Reply Coupon (IRC).

The results of RSGB contests are published in *RadCom;* if it is an overseas contest, the scores of British entrants are often given by G3XTT in the 'HF' column. If you want to receive detailed results from overseas contests, you should enclose an A4 or A5 SAE with one or two IRCs.

Duplicates

Some contests require you to submit a dupe sheet when you have made, say, 200 contacts, or more than 50 contacts on any one band. This is good practice anyway, and helpful to the adjudicator. What is required is a separate list of contacts for each band arranged in alphanumeric order with your contact number shown for each call. These lists immediately highlight the duplicated contacts.

As your experience increases it will be found that it is not difficult to make 100-plus contacts on a single band in one of the busier contests, and that compiling dupe sheets is a time-consuming, and – if several hundred callsigns are involved – a very tedious business.

Maintaining a dupe sheet as you go along is the answer to this problem when logging on paper. It requires a large sheet of paper, say A3, for each band, ruled into 26 box sections, one for each letter of the alphabet. The calls logged are recorded on the dupe sheet by suffix. For example, G2ABC would be entered under 'A' and HG65N would be entered under 'N'.

This method more or less spreads the information evenly over the sheet, making it quicker to find a given station. Listing under prefix would result in a long list of stations under letters such as D, G or W, and you would quickly run out of space in those sections, whilst the other boxes remained almost empty.

WEEKEND PROJECTS

Handling the dupe sheet as a single operator, unless it is a very slow contest, is always a difficult task, seldom fully managed. If enjoying a good run, with plenty of callers coming at you, do you work them at three or more per minute or slow it down to half that rate while you attend to the dupe sheet? You must also keep a multiplier sheet up to date at all costs. I must confess in my own case the dupe sheet is put aside, and brought up to date as best I can during slow periods. They are seldom completed satisfactorily, for a good run can soon put a couple of hundred contacts in the log, which takes some catching up on, and is also not the best use of time when there are more stations to be worked and other bands to be checked. These shortcomings are the reason why the large majority of contesters now use computer logging.

COMPUTER LOGGING

It is not many years since we were all logging on paper. In an all-band contest we required log sheets, dupe sheets and multiplier lists for each band – so much paper that it took up more room than the equipment! The amount of paperwork after a contest finished did in fact deter many people from submitting an entry. A real breakthrough came with the advent of real-time computer logging, which gives instantaneous warning of duplicate contacts and even scores the entry for you.

The logging programs are now so sophisticated it is a real joy to contest using these tools. The score is computed as the contest progresses, giving an up-to-the minute picture, and there are windows which detail which mults have been worked and which are still needed on each band.

There are many programs available, ranging from the very simple to the ultimate. The latter type includes *CT* by K1EA, which has everything including computer control of the rig, support for the Packet Cluster, networking facilities for multi-op stations and even a 'band map' of the band in use which details stations worked and highlights packet spots still needed.

Probably the most suitable program for participants in RSGB contests is SD (Super Duper) by Paul O'Kane, EI5DI, because all RSGB contests are supported. It is very user-friendly too – important if your keyboard skills are not highly developed.

ADJUDICATION

After submission, the logs are checked by the contest adjudicator. His task is to compile a database of all the callsigns appearing in the submitted logs and then check each individual log for accuracy. Special attention is given to any call which appears in one log but not in any others. Such calls may be miscopied in which case they are known as *broken* or *busted* calls, and all claimed points are deducted for that contact.

If that contact happened to also be a multiplier, another contact with the same multiplier will be looked for, so it is a good idea to include 'spare' multipliers on the multiplier list for this purpose. If the call differs greatly and cannot be regarded as broken, then it is referred to as a *unique call*. Your log will be cross-checked against others, and it often happens that your call may not appear in the other log. Even though you *think* you worked him, he was in fact contacting someone else whom you could not hear. This type of error will decrease with experience but can still occur. Even top DXers get some "not in log" responses to QSL card requests. Errors in copying any of exchange data is penalised by the deduction of one third of the QSO points per error.

Mention has been made of the punitive penalties applied in the case of unmarked dupes but an excessive number of un-removed dupes or unverifiable contacts can lead to disqualification. It is clear that entries need to be both clear and accurate.

Many errors occur during the transfer from original sheets to entry sheets, due to misreading handwriting or careless typing. In the case of RSGB contests, the adjudicators often comment on their amazement at how some operators will put in 24 hours or more hard work on the contest, but not even glance over the entry for errors after preparation before submitting it.

FINALE
It has been said that more can be learned about propagation during a full weekend contest than in a full 12 months of casual operating. You will discover the openings on the separate bands to the various areas of the world which occur at different times on a range of frequencies. Knowing when and where to look will give you an edge over many of your fellow operators when it comes to looking for DX in casual operating sessions or chasing that rare station in pile-ups.

The improvements to your station as you strive to improve your position and technique will come along gradually as you change, hone and fine-tune the aerials, equipment and layout of the station.

Most of all, you should obtain enjoyment and satisfaction from contesting. You can get a real buzz from a good event because, above all, contesting is about having fun from your hobby.

IOTA – A BEGINNERS' GUIDE

The arrival of many newcomers on the HF bands following changes to the licensing requirements makes it a good time to explain the basics of the RSGB IOTA Programme and the new world this opens up on operating.

Among programmes that stimulate daily activity on the HF bands, two stand head and shoulders above the others: DXCC for working countries, or 'entities' to use current terminology, and IOTA, for contacting island groups. The programmes are similar in character: both are international in coverage, both have a strong rule structure and neither is open-ended. Moreover, in practical terms they complement and strengthen each other because activity to promote one often provides valid contacts for the other.

IOTA, or the 'Islands On The Air Programme' to give it its full title, continues to grow in popularity each year, not only among the ever-increasing numbers of island chasers, but also among a rapidly expanding band of amateurs attracted by the possibilities for operating portable from islands. For both, it is a pleasurable pastime adding much enjoyment to on the air activity. The basic building block for IOTA is the *IOTA Group*. The oceans' islands have been corralled into some 1200 IOTA Groups with, for reasons of geography, varying numbers of '*counters*', ie qualifying islands, in each. Only in very few cases do the rules of IOTA allow single islands to count separately, DXCC island entities, such as Barbados, being one. The number of Groups is now capped and further changes are expected to be minimal.

Each Group activated has been issued with an IOTA reference number, for example EU-005 for Great Britain. Part of the fun of IOTA is that it is an evolving programme with new Groups being activated for the first time. Currently some 1050 of the 1200 Groups have numbers. The objective, for the island chaser, is to make radio contact with at least one counter in as many of these Groups as possible and, for the DXpeditioner, to provide such island contacts. A wide range of separate certificates, graded in difficulty, is currently available for island chasers as well as two prestigious awards for high achievement. Applicants may be any licensed radio amateur (or SWL on a 'heard' basis) who has had confirmed contacts with the required number of IOTA Groups listed.

IOTA DIRECTORY
The IOTA 'Bible' is the *IOTA Directory*. This gives a far more detailed description of the Programme than is possible here. It also provides a

IOTA – A BEGINNERS' GUIDE

full listing of the 1200 IOTA Groups together with the names of 15,000 qualifying islands. If you decide to participate in the Programme, either as an award applicant or DXpeditioner, or even as a 'closet' follower of the Programme, you will need to have access to a copy of either *IOTA Directory 2000* or the latest edition [1]. Earlier editions do not include the significant changes made to the rules and the island listings in year 2000, so reliance on one of these is not advised. You can obtain the latest updated information from either the IOTA Manager's or the RSGB IOTA websites.

APPLYING FOR AN AWARD
Award applicants may submit their applications electronically – for this, the preferred method, you need to obtain an IOTA Members Application Disc (*IOTAMEM*) from your Checkpoint – or on paper. Full details of the application process and a list of Checkpoints can be found in the *Directory* or on either of the two IOTA websites. When you have prepared your application, you should send it with your cards and the appropriate checking fee to your Checkpoint.

GETTING STARTED
New Licensees: You have everything to work! Follow the guidance in the *Directory* and you should be well on your way during your first year of activity. Remember you need QSLs for IOTA, so make sure to apply for them promptly.

Previous VHF-only Licensees: If you have been active for many years on VHF, you have probably already worked a large number of island stations that count towards the initial IOTA 100 Islands of the World Certificate (All Band). You may find that after just a few weeks activity on HF you can complete the 100 IOTA Groups in seven continents requirement to achieve entry.

Old-Timers: Yes, we have heard the excuses "I don't have time at present", "I will get round to it one day", or "I don't like sending cards away". The only excuses we readily accept are "I don't collect cards", "I don't collect certificates" or "I don't have any money"! If the idea of searching through countless shoe-boxes of QSLs to find a card from every Norwegian or French IOTA is too discouraging, concentrate on the 100+ DXCC island entities that comprise just one IOTA group and then the 30 extras that cover two IOTAs. These are all listed, just for you, in the *Directory*! This will at least get you started. When you are really hooked, looking through those shoe-boxes will not present such a turn-off!

Once you have a record on the IOTA database, your score is entered into the Annual Listings published

The QSL of the XF2IH operation from Enmedio Island, a 'new one' for IOTA (NA-224).

WEEKEND PROJECTS

each year in the *RSGB Yearbook* [2]. You will remain listed so long as you update at least once every five years. In fact we find that many IOTA enthusiasts are just as interested in participating in these listings as in collecting the certificates!

ISLAND-CHASING

A thousand or more IOTA Groups may seem an enormous target. If you are a long-time DXer who has worked it all and are looking for something new, you will already have amassed a very respectable IOTA score from among your DXCC contacts. If, however, you are new to the bands or one of the many amateurs who adopt a more relaxed approach to their operating, you can take full advantage of a very high level of IOTA activity, comprising easy and semi-rare Groups, to launch you on your way. Well over 700 IOTA Groups are usually activated over a three-year period with, during a typical summer weekend, some 20 – 25 IOTA Groups being heard around the IOTA meeting frequencies. An enthusiast should be able to gain the IOTA Plaque of Excellence for working 750 Groups in about five years, operating mainly at weekends. This must be a reasonable target to go for – after all, how long does it take to get to the top of the DXCC Honor Roll?

OPERATING FROM AN ISLAND

Many amateurs are fortunate enough to live on an island and to be able to give out an IOTA every time they make a contact. Others are not so lucky. For both there is the lure of operating portable from a rare or rarer Group – the fun of being for a few days at the other end of a pile-up. Many islands lie within a few hours' reach and, subject to the availability of suitable equipment, could be put on the air relatively easily. Those amateurs lucky enough to be able to activate a rare or semi-rare IOTA Group can expect to generate huge pile-ups with thousands of contacts during even a short two to three day period. Rare Groups are not all remote and difficult to access. Even in Europe and North America there are many such that are needed by the chasers. For those interested, a list of most wanted IOTA Groups in each continent, ranked by rarity, can be reviewed on the RSGB IOTA website. Thanks to the generosity of Yaesu, the principal sponsor of IOTA, the RSGB IOTA Committee has a number of portable stations that comprise a small Yaesu transceiver, lightweight switched mode power supply, microphone, keyer, and wire antenna, all boxed in a small splash-proof case. These stations can, subject to availability and a few very straightforward conditions, be loaned to anyone wishing to activate an island. Both the IOTA Committee and Yaesu are keen to introduce younger amateurs to DXpeditioning, so younger teams, particularly those from radio clubs, will get priority. Anyone wishing to borrow a portable station should contact Neville Cheadle, G3NUG, at g3nug@btinternet.com

IOTA ACTIVITY

IOTA stations tend to operate around the nominated meeting frequencies of 3755, 7055, 14,260, 18,128, 21,260, 24,950, 28,460 and 28,560kHz on SSB and 3530, 10,115, 14,040, 18,098, 21,040, 24,920 and 28,040kHz on CW.

IOTA – A BEGINNERS' GUIDE

No specific frequency has been nominated for 7MHz CW but it is recommended that operations should include a frequency above 7025kHz when the band is open to North America. Two main sources of upcoming IOTA activity are the RSGB IOTA website and the *425 DX News*, a weekly bulletin circulated by e-mail (for details on how to subscribe, see their website). These are not the only ones since, reflecting the Programme's popularity, many other websites have been set up with an IOTA focus – for a list, see the *Directory*. The big event of the year for IOTA enthusiasts is the *RSGB IOTA Contest*, held on the last full weekend of July. This provides a great opportunity to work large numbers of rare and semi-rare IOTA Groups. Mark it in your diary and join in. In fact, why not go on your own IOTA DXpedition?

WEBSITES
RSGB IOTA   www.rsgbiota.org
IOTA Manager   www.g3kma.dsl.pipex.com
RSGB HF Contests   www.rsgbhfcc.org
425 DX News   www.425dxn.org

REFERENCES
[1]   *IOTA Directory* latest edition, available from RSGB Sales.
[2]   *RSGB Yearbook* latest edition, available from RSGB Sales.

RADIO-FREQUENCY MIXING EXPLAINED

> Mixers find widespread use in electronic circuitry. Many of the projects in this book, together with every TV set and radio in the home, contain mixer circuits – a good indication of their usefulness.

CONFUSED?
Audio mixers (as used in recording studios and radio broadcast stations) are used to add or 'balance' the signals from various sources such as microphones, CD players, etc. These have nothing whatsoever to do with radio-frequency (RF) mixers, and should never be confused with them.

RF MIXERS AND BEAT FREQUENCIES
Instead of adding signals (as in the audio mixer), the RF mixer multiplies them together. As you might expect, this has an entirely different effect. The two signals entering the mixer beat or heterodyne with each other to produce signals on other frequencies. One example of this occurs in sound, when two musical notes of almost the same frequency are heard together.

Instead of hearing two separate notes, the listener hears one note whose intensity (loudness) appears to increase and decrease. This intensity variation is called a beat, and its frequency is equal to the difference in frequency of the two original notes. The technique is used by musicians to tune their instruments. If one note is known to be a correct frequency, the other can be tuned to it by making the beat frequency as close to zero as is possible. The mixing of the two original notes into what is actually heard takes place in the mechanics of the ear.

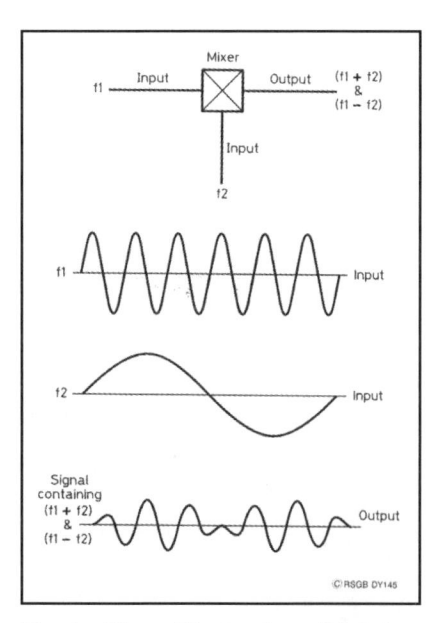

Fig 1. The effect of multiplying (or mixing) two signals together.

MULTIPLYING TOGETHER
The process of mixing presupposes that we have a device which will automatically multiply two signals together. Fortunately, this is easy; so easy, in fact, that it often occurs when we do not want it! Multiplying is achieved by any device which is non-linear; this means a device whose output is not a constant factor larger than its input, something that can be achieved by many electronic devices and circuits.

RADIO-FREQUENCY MIXING EXPLAINED

Let us look now at what a mixer does in concrete terms. Suppose two signals, of frequencies $f1$ and $f2$ go into our mixer. These signals are shown in Fig 1. Putting numbers in, to make the situation clearer, Suppose $f1$ is 1.000MHz and $f2$ is 160kHz. The beat frequency is the *difference* of these:

1.000MHz - 0.160MHz = 0.840MHz, or 840kHz.

A mixer also produces an output at the *sum* of these frequencies; in this case the new frequency would be:

1.000MHz + 0.160MHz = 1.160MHz.
i
Suppose you fed the output of your mixer, operating with these input frequencies, into a receiver and tuned around to find what frequencies were present. You would find two signals, one at 840kHz and one at 1.160 MHz, showing that the two 'new' frequencies were very real!

In addition to drawing out the waveform of the resultant signal, as in Fig 1, we can draw the inputs and outputs on a frequency axis, to form a spectrum of the signal components. This is done in Fig 2. The top two diagrams show the input signals at $f1$ and $f2$. The bottom diagram shows the output signals in relation to the input signals. Depending on the type of mixer used, one or both of the input signals would subsequently be removed.

A MIXER IN EVERY RADIO

Basically, a mixer is used to change a signal from one frequency to another, something it does without altering the characteristics of the incoming signal. If the incoming signal is amplitude modulated (AM), then the frequency-changed signal would be AM also. The same applies to FM, SSB, CW, and all other modulation forms you can think of. This explains why mixers are often called frequency changers.

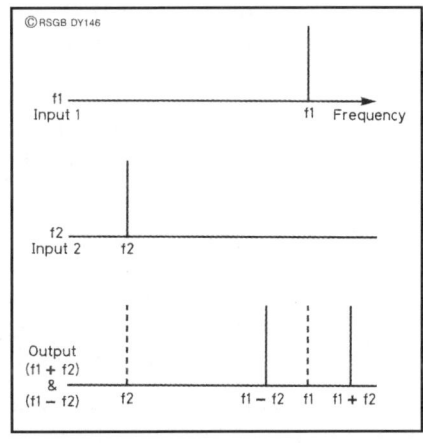

Fig 2. The result of mixing two signals together as seen on a spectrum analyser.

Frequency-changing is the key process in the type of radio known as a superheterodyne (or superhet). By mixing the incoming signal with a variable-frequency local oscillator, as Fig 3, shows, the signal can be converted to the fixed frequency of a filter and amplifier. This is useful because it is easier to make a very high-quality filter and amplifier at a single fixed frequency, than at a variable frequency. All TV receivers and virtually all radio receivers (and transmitters) use mixers.

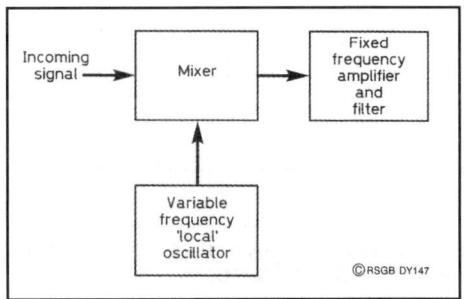

Fig 3. The basic idea of a superhet receiver.

209

NOISE-REDUCTION CIRCUITS

Many people have transceivers with digital signal processing (DSP) inside them. Here is a simpler approach, based on an analogue integrated circuit. The prototype used some surface-mount devices (SMDs) but this is not necessary to enjoy the benefits of this little circuit.

DSP
The test reports on digital signal processors promise miracles; many people feel that the noise is reduced in the noise reduction mode but so is the readability!

The Analog Devices SSM2000 IC [1], works on the HUSH principle, developed and patented by Rocktron Corporation. Although designed for the hi-fi market, it has found an application in amateur radio which is described here.

HOW IT WORKS
The HUSH system can distinguish between signal and noise because the volume and frequency spectrum of speech or music continually changes; by contrast, the noise amplitude and frequency spectra remain relatively constant.

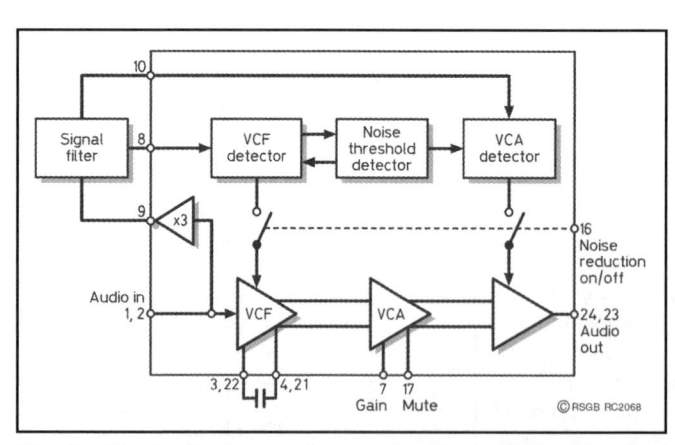

Fig 1. Block diagram of one channel of the Analog Devices SSM2000 IC.

Following the block diagram in Fig 1, the audio signals (only one channel of the two required for stereo sound applications is described) are processed to extract information concerning the frequency distribution and amplitude of both signal and noise, passing through low-pass voltage-controlled filters (VCFs) and then through voltage-controlled amplifiers (VCAs). In contrast to digital processes, VCFs and VCAs have low distortion and add negligible noise of their own. The cut-off of the VCF and the gain of the VCA can be set as required by the application. With control signals derived by a proprietary algorithm and applied to both VCF and VCA, a noise reduction of up to 25dB can be achieved.

NOISE-REDUCTION CIRCUITS

THE APPLICATION

Looking at the diagram in Fig 2, the signal is applied to pins 1 and 2. It comes out of pin 9, amplified by a factor of three, from where it is fed unaltered to the VCA detector (pin 10) and through a three-pole high-pass filter to the VCF detector (pin 8). Fig 3 shows the two frequency responses, optimised here for hi-fi music; for narrow-band voice signals it would be appropriate to move the frequency response of the three-pole filter downward by making the capacitors about five times bigger (22nF to 100nF, 22nF to 100nF and 2.2nF to 10nF respectively). The time constant for the decay of the control voltage to the VCFs is set by the 1µF capacitor on pin 11. The noise threshold is 'held' on the 220nF capacitor on pin 15. The control voltage to the VCA decays according to the value of the capacitor on pin 12. The lower cut-off frequencies of the VCFs are determined by the 1.5nF capacitors between pins 3 and 4, 21 and 22.

The supply voltage on pin 5 is decoupled by the BC548 emitter follower. The op-amp holds pins 6, 7 and 14 on half the supply voltage. The switch on pin 16 allows the noise suppression to be disabled.

RESULTS

The unit was built on a double-sided 52 by 33mm PCB. Except for the DIL ICs, SMD components were used throughout. Through-connections are by way of the terminal pins and the pins of the DIL sockets, which are soldered on both sides. The two 1.5nF capacitors were soldered directly to the IC.

There are no adjustments. As an audio input of 0.3V is recommended, the noise suppressor was installed between the product detector

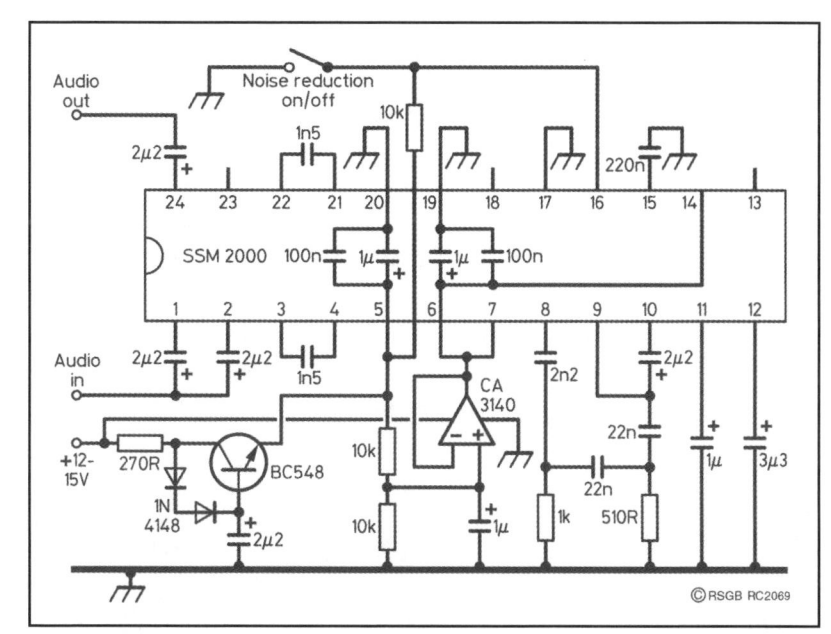

Fig 2. The noise reduction circuit, for use in a communications receiver.

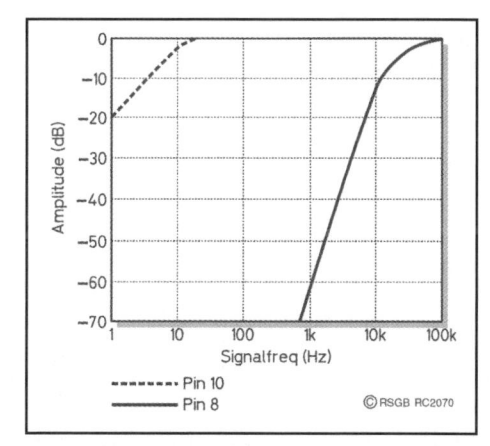

Fig 3. Low-frequency roll-off to pin 8 through the 3-pole filter in Fig 3. Also shown is the signal on pin 10, which does not go through the filter.

WEEKEND PROJECTS

and the volume control of the receiver, an old Drake R4-C. The product detector output is fed to both the L and R inputs of the SSM2000. The sum of these inputs may be applied to pin 9 of the VCA and VCF controllers. The output to the volume control is taken from one channel only.

The manufacturer claims a noise reduction of up to 25dB in hi-fi systems. This cannot be obtained with speech, but the performance is very pleasing as it is achieved without audible distortion of the wanted signal.

REFERENCES
[1]   A 16-page data sheet for the SSM2000 can be downloaded from www.analog.com.

THE PHOTOMETER AND THE POLAR DIAGRAM

Before the days of automatic 'point-and-shoot' cameras, a photographer would use a exposure meter (or photometer) to measure the light level, then manually convert this reading into shutter speed and lens aperture settings to ensure a correctly exposed negative. Modern cameras have quite sophisticated photometers, which control the shutter speed and iris settings automatically.

A SHORT EXPLANATION

A simple photometer circuit is shown in Fig 1, and is based on a device called a light-dependent resistor, or LDR. As its name tells us, its resistance depends upon the amount of light falling on it. In bright light, the resistance is relatively low (about 1kΩ), whereas in the dark, its resistance is very high (up to 1MΩ). The cell is made from a semiconducting substance known as cadmium sulphide (CdS), and is enclosed in a small plastic container. The semiconductor is laid on a flat insulating surface in the form of a small flat ribbon. The ribbon construction gives a good area of surface for a given length of ribbon, and the length of the ribbon is maximised by laying it out in a zig-zag pattern. In the dark, CdS is an insulator; when light falls on it, electrons are released inside the CdS, making it conduct. The more light there is, the more electrons there are, and the resistance falls.

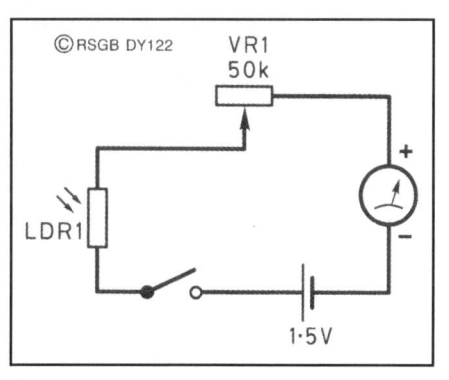

Fig 1. Circuit diagram of the photometer.

THE PHOTOMETER CIRCUIT

The circuit is simply a series connection of four things – the battery, the LDR, a variable resistor and a meter. A switch is also provided. The combination of the LDR resistance and that of VR1 determines the current flowing, which is indicated on the meter. Altering the resistance of VRI sets the sensitivity of the photometer – you may want full-scale deflection of the meter needle for a bright light, or for a dim light.

It is simple to make, and a plug-in type matrix board is ideal to test the circuit, so that you can decide if you want to make a permanent version. Connect all the components in series; the only change you may want to make is the connection to VR1. If you find that the sensitivity control seems to work 'backwards', simply unsolder the wire from the end tag of VR1 and solder it to the opposite end tag. Problem solved!

WEEKEND PROJECTS

IN USE

As soon as you connect up the battery, you will probably have a meter reading because of the daylight falling on the LDR. Shading it with your hand should reduce the reading. If the meter needle is hard over against the end-stop, turn VR1 until it indicates about half-scale. The LDR is very sensitive, and will read zero only in a dark room. If you put on a torch to see what the meter reading is, the LDR will detect the torch light, and will give a reading!

Here is a simple project where you can use the photometer in an experiment which has an analogy in radio. Draw a circle on a large (A3 or bigger) sheet of paper and divide it up into 30° sectors, as shown in Fig 2. Draw a series of smaller circles which divide the maximum radius into five. Look at the figure if you're not sure about this. Bring the LDR away from the circuit by using two long, flexible wires. The experiment must be performed in a darkened room (preferably in total darkness). Prepare a table with two columns, the left-hand one headed 'Angle (degrees)' and the right-hand one 'Meter reading'. Fill in the left-hand column 0, 30, 60 ...and so on up to 360°.

Place the torch in the position shown, with its lens at the centre of the circle and pointing along the 0° line; switch it on. Place the LDR facing the torch and adjust VR1 until you have full-scale deflection of the meter needle. Suppose the meter indicates 10 units at this point. Enter this into your table in the 0° row. Keeping the torch the same distance from the circle centre, and pointing at it, move the LDR round all 30° positions and record the meter readings. Switch off the torch and take the sheet of paper and your tabulated results into daylight!

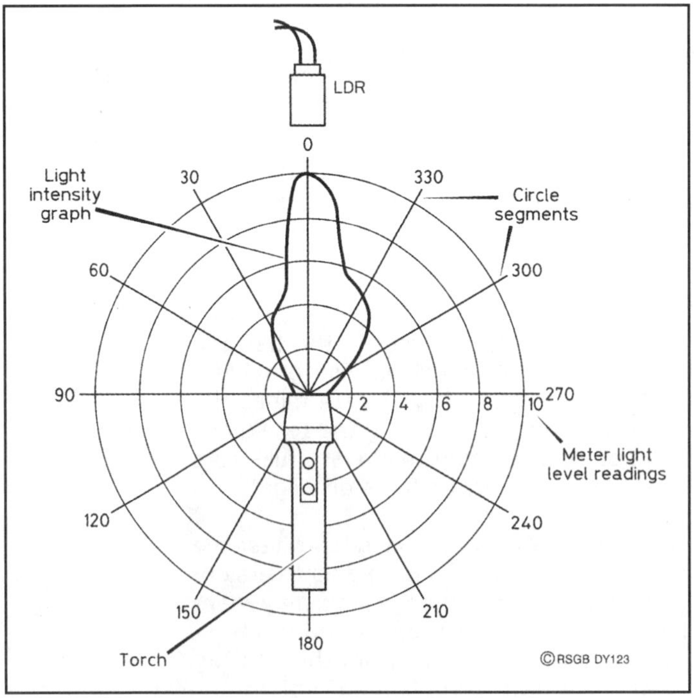

Lay the large sheet of paper on a table with your results beside it; then, at each 30° interval, plot the point along the radius corresponding to the meter reading. Then, join up all the points and you have what is called a 'polar diagram' of light intensity. The use of the word 'polar' implies that the readings have been taken in a circle and plotted that way.

Fig 2. A typical torchlight intensity pattern.

Light waves and radio waves are both examples

THE PHOTOMETER AND THE POLAR DIAGRAM

of electromagnetic radiation; only the frequencies are different. The torch is designed to 'beam' its radiation in a particular direction, just like an aerial does. Hence the use of the word 'beam' for a directional aerial. If a similar polar diagram is drawn for a Yagi-type aerial, it will show the same general characteristics as does Fig 2, namely a main direction (or 'lobe') where most of the energy is concentrated, with evidence of sidelobes, indicated by 'lumps' on the sides of the otherwise smooth main lobe.

PARTS LIST
Resistor
VR1   50kΩ linear

Semiconductor
LDR1   ORP12

Additional items
Meter   50 or 100µA full-scale deflection (FSD)
             Battery clip for single AA cell
             AA cell

THE QSL BUREAU SUB-MANAGER'S TALE

Many newcomers to amateur radio want to start a QSL collection as soon as they get on the air and, with the influx of former Class-B amateurs on the HF bands, many of whom will also want to start collecting QSLs, there is no better time for a reminder about the workings of the RSGB QSL system. Graham Ridgeway, a volunteer QSL Sub-Manager of many years' standing, explains what you can do to help make the system run smoothly.

I am aware that every amateur and SWL in the UK knows – because they have read the relevant section in the *RSGB Yearbook* – the basics of sending and receiving QSL cards. What do you mean, you haven't and have no need to? Perhaps you should. You might discover that all outgoing cards are sent to the RSGB, and *not* to your Sub-Manager (oh yes, it happens!) Also that pre-printed address labels are available upon receipt of an SASE, from RSGB HQ. [*See QSL note below*]

IMPROVING RETURN RATES

So, how does one improve the return rate of cards? Have you supplied your Sub-Manager with envelopes? No? Perhaps that's why your return rate is so abysmal. Or have you not restocked his supply? He is unlikely to chase after you. This is why you are asked to number all envelopes, and mark one clearly as "Last Envelope", so that *you* know when to restock. Your Sub-Manager is, please remember, a volunteer, doing this task to help you. You do not have to be an RSGB member in order to *collect* cards (although it would be preferred of course) but you *do* have to be a member in order to *send* cards out via the bureau. But just because you have not sent any cards, do not assume that none have come in for you.

Photo 1. Graham Ridgeway, M5AAV, sorting cards from overseas for M5 licensees.

Right, you now have envelopes waiting for those cards to arrive through the system. These are, one would hope, nice manilla C5 size, and you have used 'non-monetary' stamps – ie marked '1st' or '2nd' class, so that nobody has to worry about Royal Mail price increases. Oh, and you did get the envelope in which you sent them to your Sub-Manager weighed at the Post Office, and not just stuck any old stamp on it? Sub-Managers do not like forking out the thick end of a pound for under-stamped envelopes, even though you would reimburse him as soon as asked, wouldn't you?

THE QSL BUREAU SUB-MANAGER'S TALE

It is of course appreciated that not everyone wants to collect cards. So why not, instead of just ignoring the fact that your Sub-Manager will have to store these unwanted cards for at least three months, do the courteous thing? Drop him a line or e-mail and tell him. At least then he knows where he stands, and can file any cards accordingly upon receipt. And please do not say, "I don't know who my Sub-Manager is." He is listed in the *Yearbook*, on the RSGB website, on my website, or a quick telephone call to RSGB HQ will provide the information required. [*See QSL note below*]

Another thing to remember is that when you go on holiday, or on that 'mini-DXpedition', which involves the use of another UK prefix, that sub-manager will *also* require to be supplied with envelopes. In other words if you as, say, an M3, operate from Scotland as an MM3, some envelopes should be lodged with the MM3 Sub-Manager. The same applies when you change your callsign. Instead of just forgetting to send envelopes to your old Sub-Manager, let him know your new call, and he may redirect them for you. Even better, make sure he is still supplied with envelopes which will keep you going for a while: even though you stop using a certain callsign there will still be cards arriving which can take quite a while to get through the system.

Photo 2. Boxes for despatching cards to the Sub-Managers.

THE SYSTEM

One question I am frequently asked is either, "How long does it take to get a card back through the system?", or "Why does it take so long?" The best explanation is to try to explain how the bureau system works.

You write out your cards, pack them up and send them to RSGB. These are then sorted into the relevant sections waiting for despatch to the overseas bureaux. Once a suitable weight is attained, the cards are sent off. Upon receipt at an overseas bureau, these boxes are opened (how soon after receipt we shall never know) and sorted for despatch to the relevant Sub-Manager at the other end. Again there is a wait for a certain weight or quantity to be reached before they are forwarded. One must now bear in mind that this overseas Sub-Manager is also probably a volunteer, and may not sort his cards as soon as they arrive. One will then assume that he has envelopes for the amateur to which your card is addressed, and again, there could be a delay until a certain weight limit is reached (because we Sub-Managers wait for an envelope to be filled unless you have stated on your envelope 'Wait 6' or 'send whenever'). Eventually that amateur will receive the cards, check his log, and hopefully write you a return card. Then, the whole procedure is reversed.

A year to make the round trip is quite good, two years not exceptional, three or more certainly not unknown. This is always assuming that

WEEKEND PROJECTS

the overseas amateur has envelopes lodged with his bureau. If not, your card will fall into that great big black hole, or returned, as the DARC (German national amateur radio society), among others, is very good at doing.

WRITING YOUR CARDS
When you write your cards, make sure that all details are correct. Date, time (GMT/UTC of course), band, mode, report and callsign. Also, if one is being used, the callsign of the other station's QSL Manager, eg 'via W1XXX', should be shown nice and clearly. When you sort your cards into prefix order before sending them off (you do sort them, don't you? Oh good) – it is the *QSL Manager's* call that is used in that sort. If you have Internet access it often pays with a 'rare one' to do a quick check on **QRZ.com** to see how the station or QSL Manager wants the cards. Unfortunately, too many these days seem not to use the bureau system, and will only QSL direct, when an IRC or 'Green stamp' (US $1 bill) plus envelope is required.

Your cards are now written and packed, all the same way up, callsigns nice and legible, *and correct*. I add that with good reason: I am not aware that the M4 series, for example, has been issued, and I'm sure that M5RIC did not operate from '9E' in 2002.

While on the subject of callsigns, a word here to the listeners. Please listen for more than 10 seconds before scribbling out a card. In fact it is always good practice to report on two or three contacts, that way you will be certain you have the call correct, and the report will be more meaningful. Perhaps when designing your cards, you may like to consider incorporating this feature.

I have found that the greatest number of errors in respect of callsigns is on listeners' cards and on QSLs con-firming CW contacts. As a mode, it may get through when conditions are down, but it is also more prone to misreading/mis-sending. One dot makes a huge difference.

So your package has been sent off to HQ. You have stamped it correctly and made sure it is securely packed. Now you can sit back and wait, while working the next set of juicy DX, knowing that you have done everything you can to ensure smooth progress for your cards through the system...

...OR HAVE YOU?
There are recommendations laid down for the sizes of cards. Do yours conform? We all know about the 'army blanket' cardboard favoured by some Eastern Bloc countries, as well as the 'rice paper' so beloved by others; hideous to handle in bulk. But then so are cards which are too small (75 x 50mm) or too large (200 x 150mm). These will either get lost, folded or torn somewhere along the line – trust me. So try to keep them at around the 140 x 80mm size.

Computer-generated labels stuck on your cards are fine, and easy to read, assuming of course that they were correctly lined up when printed (I have seen too many with half the print missing, invalidating the

THE QSL BUREAU SUB-MANAGER'S TALE

QSL). Also, please do not alter anything on a card: if you make an error, write out a new one, as cards with alterations on are invalid for awards.

I am often asked whether it is worth getting multi-coloured, double-sided cards printed in an attempt to increase the 'collectability' of your cards. There is no simple answer to this. It is your choice, and depends a lot upon whether you are going to send a card for every QSO made (expensive!) or just for those for which you really wish to get one in return. Or perhaps you are someone who only responds to incoming cards. Although it may sound obvious, if you do chose the 'multi-coloured' route, just think a bit first. Ask yourself if you would be happy to receive your card. Does it stand out? Is it clear, and not over-fussy? After all, at the end of the day, you are hoping for a card in return. Your card, and the way it is presented, tends to be a reflection of your character.

Photo 3. Jan Case, RSGB HQ QSL Bureau Supervisor in 2004, sorts cards for M3 licensees.

I was asked by one of 'my' M5s why it was that although he had sent out over 300 cards he had only received six back. Explaining the torturous route of a card helped a bit; he had only been licensed a year at the time. My own return rate (three years plus) is around 15% on HF so far, and rising. On VHF, over 20 years of operating mainly on 2m produced a return rate of just 41%, although on 70cm the return rate was 83%. So you have to accept that not everyone wants or sends cards.

So, what to do if you either do not get a card back or if yours is returned marked 'Does not use the services of xxx bureau'? In the latter case, if you really want a card from that station, try the direct route. In the former, assuming a reasonable length of time has elapsed, the direct route may work, as may a second card via the bureau, although I have seen cards from overseas marked 'third request' for contacts over 10 years previously.

I would suggest that one will not be forthcoming after that period has elapsed. Details of QSL managers for a lot of DX stations can be found in the 'HF' column in *RadCom* on a regular basis, on the Internet (eg the *425 DX News* service), and also on packet radio. So, spare a thought next time you send some cards off, not only for Jan and her happy crew at the RSGB HQ, but for all the volunteers who keep the system running. Help them to help you. Your Sub-Manager is there to help if

WEEKEND PROJECTS

needed. Most can be contacted by e-mail, all by 'snail mail' (and an SASE really is mandatory), and I for one am not averse to receiving telephone calls.

With that, may I wish you good DX, and an improvement in your return

ON THE WEB
RSGB (Go to 'RSGB Information' then 'RSGB QSL Bureau')
www.rsgb.org/membersonly
G8UYD/M5AAV   www.users.zetnet.co.uk/m5aav/index.htm
QRZ.com   www.qrz.com
425 DX News www.425dxn.org

Editor's Note: as from April 2008, the Headquarters of the RSGB is moving from Potters Bar to Bedford. One of the departments affected is the QSL central bureau, and readers needing an address to which their outgoing cards should be sent are strongly advised to consult the RSGB website (www.rsgb.org), where all current QSL information can be found.

RADIATION RESISTANCE

> Radiation Resistance is a most interesting and very important phenomenon associated with aerials, but it is one that is not always fully understood. Here is a simple and straightforward approach to the subject.

Assume for example that we are testing a transmitter and the indications are that it is supplying 100W of power to a resonant half-wave aerial at a height of $^1/_2$ wavelength, fed at the centre with 72Ω flat twin cable. Appropriate steps have been taken to ensure that the balanced feeder and the unbalanced nature of the transmitter output have been catered for; the VSWR is also very low. If the aerial were replaced with a non-reactive 72Ω resistor sufficiently large to dissipate 100W, provided the frequency to which the transmitter was tuned was left unchanged, the change between aerial and resistor would be undetectable at the transmitter end of the feeder.

This is the way that the Radiation Resistance of an aerial is described, perhaps elaborated a little, but essentially the same. For example: *'The total amount of energy radiated from a transmitting aerial can be measured in terms of a Radiation Resistance which is the resistance that, when replacing the aerial at the end of the feeder will consume the same amount of power that is actually radiated'* is typical of the many text books dealing with aerials. While conveniently describing the phenomenon, the impression given might lead you to assume that the Radiation Resistance is in fact a resistor.

AN ALTERNATIVE VIEW
Looking at an aerial in a somewhat different way from that which we are accustomed, it might described as *'a region of transition between a wave guided by a transmission line and a free-space wave'*, or perhaps that *'an aerial interfaces between electrons on conductors and photons in space'* (Fig 1). Both these quotations make you realise that an aerial is something rather special and that the point of departure of our electromagnetic energy cannot be just a resistor, for if it were, the energy would be dissipated at the site of the aerial and there would be none left to journey onwards.

The complex nature of an aerial when radiating energy, and indeed when receiving it, has been investigated mathematically. It is to the mathematician that we owe the calculation of the radiation resistance of many kinds of aerial, though the dipole is the fundamental building block. For example, the radiation resistance of 'short dipoles' has been carried out and produced the following results:

WEEKEND PROJECTS

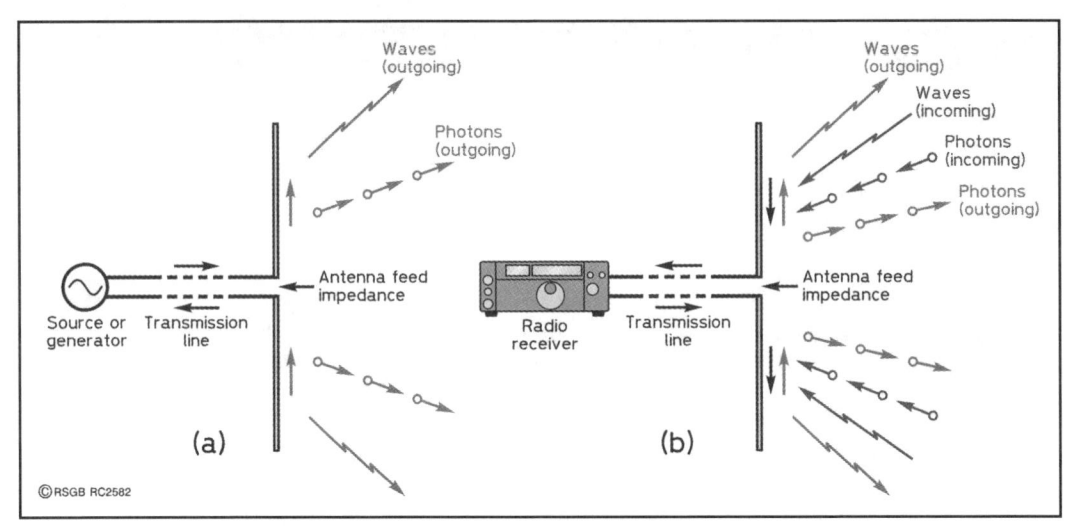

Fig 1. Energy (a) leaving, and (b) leaving and being received on a dipole antenna. A photon is the quantum unit of 'electromagnetic energy'.

Length	Radiation Resistance
$\lambda/10$	7.9Ω
$\lambda/100$	0.08Ω

The radiation resistance of a short dipole is therefore small.

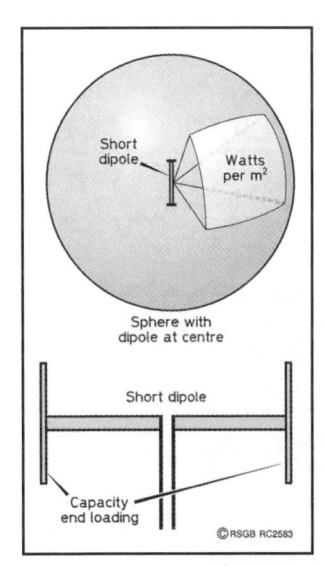

Fig 2. Radiation resistance is calculated by placing a dipole at the centre of a sphere that is large with respect to wavelength.

CALCULATIONS

To calculate radiation resistance, the dipole is considered to be at the centre of a sphere (Fig 2) that is large with respect to wavelength, and here the 'Poynting Vector' of the 'far field' (ie many wavelengths from the dipole) is used to obtain the total power radiated.

Assuming no losses, this power is equal to the power fed to the dipole. From Ohm's Law it can be calculated that the power P must be equal to the square of the RMS current flowing in the dipole, times a resistance R called the Radiation Resistance. This total power is the rate at which energy is streaming out of the sphere surrounding the dipole. For these calculations the length of the dipole is considered to be *very much less* than the wavelength used in the calculation. All very theoretical, perhaps, but nevertheless shown to be sufficiently near the truth when investigated in a practical situation.

Virtually the same procedure (although different mathematically) is carried out for a half-wave dipole, but this time the Radiation Resistance is found to be 72Ω. This is considered to be at the current maximum point on the dipole, ie the centre, which is also the point where the feeder is connected. The half-wave dipole may be utilised on *odd harmonic frequencies*, the lowest being the third harmonic where the radiation resistance will be about 90Ω, on the fifth about 120Ω - all of these at the centre of the aerial. For even harmonics the centre point of the aerial will have a high (or very high) resistance, the 72Ω feeder would be unusable and open wire line would have to be utilised. This of course describes the multi-band aerial. The popular G5RV is non-resonant, except perhaps nearly so on 20m, (about three half-waves), but on 80m the centre impedance will be something like 30Ω resistance and -500Ω reactance (ie 30Ω - j500Ω).

PRACTICALITIES

When looking at a practical aerial there will be a number of other points which must be taken into consideration, not the least of which is the environment in which the aerial is erected. The working value of the radiation resistance depends on:

- the relation of the aerial to ground;
- the ratio of conductor diameter to length;
- the proximity of other conducting objects such as masts, buildings, house wiring, telephone wires, etc.

For example, the feed-point impedance of a practical half-wave dipole is about 73Ω plus 42.5Ω of inductive reactance ('end effect'), ie 73Ω + j42.5Ω. In addition to the radiated energy, energy is also lost in the resistance of the aerial wire, the resistance of the ground, along with dielectric losses in trees, insulators, and losses in other objects with an imperfect dielectric (Fig 3). The losses are brought together and included with the natural radiation resistance when describing the 'feed impedance', which is often used in practice.

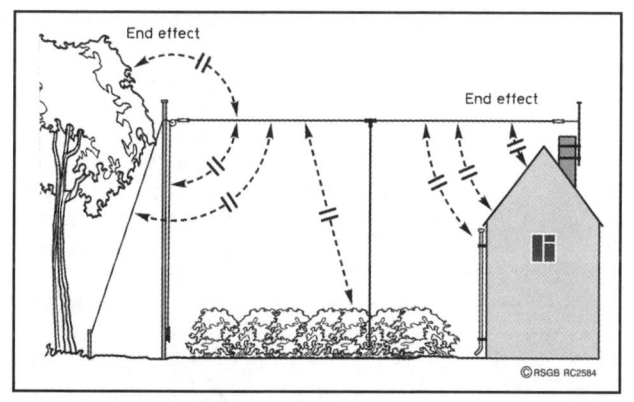

Fig 3. How radiated energy is wasted due to dielectric losses.

It is perhaps evident that the true radiation resistance of a particular aerial is not normally measurable, due to the factors mentioned above, hence the feed impedance is equal to the radiation resistance plus the loss resistance,

ie $R_{aerial} = R_{radiation} + R_{losses}$,

where R_{aerial} represents the feed impedance.

The efficiency of the system is given by:

$R_{radiation} / (R_{radiation} + R_{losses})$,

and is usually very high when dipoles and other centre-fed systems are concerned (in the region of 90%).

WEEKEND PROJECTS

Another type of half-wave aerial is the folded dipole. This aerial has a somewhat different feed impedance characteristic. A two-wire folded dipole has a feed impedance of about 280Ω, though it is generally taken as 300Ω. The feed impedance of a folded dipole varies in a non-linear fashion. For a three-wire half-wave folded dipole, the feed impedance is 630Ω. In general, for a half-wave folded dipole with 'N' wires, the feed impedance is $70N^2$Ω.

Beam aerials are a case where the proximity of reflecting and directing elements change the feed impedance substantially. For a two-element beam with 0.3λ spacing, the dipole or driven element feed impedance falls from about 72Ω to about 65Ω, due to mutual coupling. As the spacing is further decreased to the optimum of 0.11λ, the feed impedance falls to 20Ω (Fig 4). This indicates that the presence of other aerials nearby any aerial in use will most certainly have an effect on its feed impedance.

Turning from the horizontal aerial to the vertical, and in particular the quarter-wave vertical fed against ground or radials, in theory this has a radiation resistance of about 35Ω (half that of the horizontal half-wave dipole). Whilst this may well be correct, the actual feed impedance is rarely – if ever – near to 35Ω. The problem lies mainly with the conductivity of the ground beneath the aerial. Experiments have shown that a number of radials laid out surrounding the vertical will bring the feed impedance to a value that can be matched, and the aerial will take energy. The efficiency of the vertical can be rather poor, but provided sufficient radials are used (up to 40) it can be brought up to about 70%. While the vertical has low angle radiation, good for long distance communication, the nature of the vertical aggravates the mutual coupling, hence the presence of trees, foliage, etc, makes the losses greater, which of course affects the feed impedance.

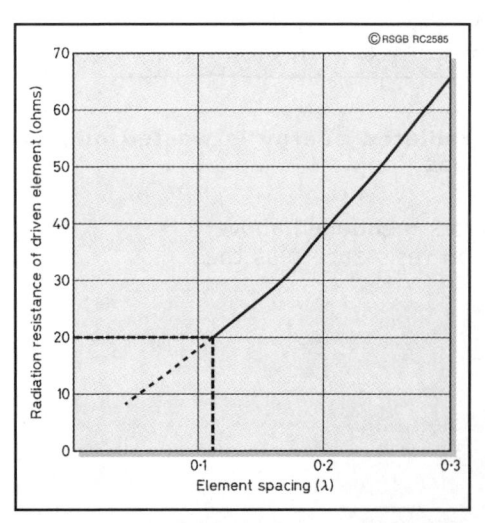

Fig 4. How the radiation resistance of the driven element of a two-element beam is affected be element spacing.

SUMMARY

There is a considerable variety of different types of aerial and as many differing radiation resistances. There is also another way that is sometimes used to describe Radiation Resistance. Some textbooks will explain that 'it all depends at which point along the aerial the current is measured and where the feeder is to be connected'.

Many aerials are end-fed and hence, like the vertical, have a return to the source via the ground. If the aerial happens to be a half-wavelength and is fed at the end, the feed impedance is going to be high. But, nevertheless, the current squared times the resistance at that point will equal the energy radiated... just the same as for the centre-fed aerial. The resistance measured is sometimes called the radiation resistance.

A BEGINNERS' GUIDE TO RTTY CONTESTS

Since the advent of the use of the computer soundcard to decode digital signals, there has been more activity on RTTY than ever before. One now hears more stations during a single day on RTTY than were normally heard in a week a few years ago. Much of this increase is due to the very effective, and free, *MMTTY* [1] program, by Mako Mori. This in itself has brought thousands of people world-wide on to the mode to enjoy the casual, friendly world of RTTY.

Many RTTY operators use RTTY contests simply as a way of increasing their DXCC totals, working towards a 'Worked All States' (WAS) or 'Worked All Zones' (WAZ) award, or picking up new prefixes and many of them continue to use *MMTTY* to this end.

RTTY CONTEST PROGRAMS
While this is all very well, and can also give the 'novice' contester a flavour of what can be worked during a RTTY contest, *MMTTY* is *not* designed as a contesting program. It therefore does not have the capabilities of a dedicated contest package such as, say, *WriteLog* [2] or *RCKRtty* [3]. These programs can recognise incoming callsigns preceded by 'DE', track your score, show needed calls, multipliers and duplicate contacts and do much more to help you maximise your score.

While *MMTTY* will not always produce a contest log in the format specified in the rules (particularly if a 'Cabrillo' log is required), all specialised RTTY contesting programs, like the two mentioned, will produce the correct format needed when sending in your log. A 'Cabrillo' log, by the way, is a standardised log format in which all the information is in one file that includes the contest log, your name and address, callsign and comments.

An alternative for producing logs is *Cabrillo Tools* by WT4I [4] which will convert various files into the correct Cabrillo format for you. This type of log is now required in all RSGB HF contests and most RTTY contests. Most logs can be e-mailed to the contest address and will be acknowledged. Any comments you have about the contest can be included in the 'soapbox' section of the Cabrillo log, but not in your e-mail text, as this may only be seen by a 'robot' mail handler!

Always send in your log, no matter how many or how few contacts you made. Your log, yes *yours*, with only 20 or so QSOs in it, is used to verify the points claimed by the other contestants whom *you* worked. "But with only 20 QSOs, I'll come last", I hear you say. I doubt it! While trying out some new software, Dick, G3URA, ended up working

WEEKEND PROJECTS

just 16 stations in one contest and when the results came out, there were at least five stations below him. Despite a 'low' score there is also the possibility that you may be the *only* G, M or 2E station to submit a log in any particular class, meaning you could end up with a certificate for being the top G in that class. Before submitting your log, do read through it and make sure it looks OK. Ensure there are no obviously wrong calls or daft exchanges. Printing the log out and then looking at it may help you spot anything untoward.

BEFORE THE CONTEST
Entering a contest can be a bit daunting the first time. The following few paragraphs are aimed to help the beginner take the plunge and offer a few tips from both avid contesters and a contest manager. The first rule in any contest must be: 'Read The Rules!' The rules for all major RTTY contests can be found on the web [5], and are also published in the British Amateur Radio Teledata Group (BARTG) [6] monthly magazine, *Datacom*, together with detailed operating tips for all the major contests.

The rules will state date and time, the exchange required, where to send logs, and in which format they must be. The rules will say if there are different classes of entry, and a single band entry may suit you if you are restricted by time or aerial considerations.

Always keep a copy of the rules handy. Before the contest, set up some simple memory 'buffers' containing only the minimum information required for calling another station and exchanging the relevant information. It is also a good idea to add a couple of 'return' characters to the beginning and end of the exchange, as this can make your exchange stand out a little more.

Phil Cooper, GU0SUP, operating RTTY from Guernsey.

For a contest where the time also has to be exchanged, the buffers might look like this:
'DE GW4SKA GW4SKA K' (Use this to answer a CQ. Never send his call and always send the 'DE', see below).

'RGR UR 599 001 001 1254 1254 DE GW4SKA K' (meaning 'I have your message; this is mine for you').

These will be fine in most conditions but be prepared to repeat the serial / time etc several times if copy is poor, like this:

A BEGINNERS' GUIDE TO RTTY CONTESTS

'001 001 001 TIME 1254 1254 1254 QSL? BK' Set up a separate buffer for this. There is no need to repeat the RST as it is always 599 no matter what the conditions!

DURING THE CONTEST
Never send any unnecessary information such as names, your rig, power or aerial details. Also, even if the station worked is a new country for you, never ask for his QSL information, as you can find this out after the contest. Remember that the serious contester will be aiming to make about two contacts every minute, so stick to the essential information only. Remember too, to call exactly on the other station's frequency and keep the 'AFC' and 'NET' controls turned *off* when answering a CQ.

Think about your exchanges and watch what others are doing. Most of us know our own callsign so seeing it three times before we see your callsign just once, is a real 'no-no'. For example, sending "GU0SUP GU0SUP GU0SUP DE G3URA PSE K" will probably not get you that much-needed GU multiplier!

Know what the 'multipliers' are. Are they countries? Prefixes? Zones? This will be explained in the rules, which you will have read before the contest, right? Are there bonus points for working different continents? If you run a 'little pistol' station where anything outside Europe is a bonus, don't forget that the Canary Islands, EA8, counts as Africa, and Cyprus, 5B4 or ZC4, counts as Asia. Both are fairly 'local' and easily worked.

Having a 'little pistol' station can have some advantages, as most of the time you will be in 'S & P' ('search and pounce') mode while the 'big guns' will sitting on one frequency calling CQ. You can pick and choose whom you work; they can't. Never forget that they *want* your call and will do their best to get you in their log. This is especially true if you have a regional locator in your callsign, such as GM or GI.

If they are rare DX and have a huge pile-up, worry not. If it is a 48-hour contest wait 24 hours and call them on the second day, when they will be crying out for contacts and will want you in their log. If you do have to wait patiently for your turn to work the DX, again, watch what is going on: there should be no need to ask for a repeat of his serial number, for example.

In some contests, such as the Australian ANARTS, points are based on distance worked and in these types of test it is far better to trawl the bands looking for DX rather than just work mainly European stations.

Watch out for time limits on band changes or off times. These will be in the rules. If, for example, you are limited to two band changes in 10 minutes don't work that one multiplier on 10m if you can't hear any other stations, or you will then have to sit on a quiet band until the 10-minute time period has elapsed (however, if you just can't resist working that VP6 before returning to 20m, you can always use up the rest of the time with a 'comfort break'!)

WEEKEND PROJECTS

If you are keen to try CQing, even with your 'little pistol' station, think about doing so in the dying hours of the contest. Then, many of the big boys will start to search and pounce for those extra contacts that escaped them during the main part of the contest.

INTERESTED?

More information about RTTY contesting and the datamodes in general can be found by joining BARTG [6], from which several RTTY awards are available. There is also a popular RTTY reflector [7] where you can ask questions, find QSL routes, and compare contest scores. Other helpful information can be found on the 'RTTY Info' website [8], where there is an excellent RTTY tutorial for those wishing to learn more about the mode.

New RTTY operators will find contesting a very easy way to make a start on the mode without the need to type at furious speeds. Those with more experience will know that in any of the major contests held each year, they can find well over 1000 stations to work. Look at the contest calendar [5], read the rules, join in, but most of all, have *fun*!

We look forward to seeing you on our screens and seeing your calls listed in the results. Oh, and one last bit of advice: Read the rules - *again*!

REFERENCES

[1]   www.qsl.net/hammsoft
[2]   www.writelog.com
[3]   www.rckrtty.de
[4]   www.wt4i.com
[5]   www.rttyjournal.com/contests
[6]   www.bartg.demon.co.uk
[7]   http://lists.contesting.com/mailman/listinfo/rtty
[8]   www.rttyinfo.net

SAFETY, OPERATING PRACTICE AND THE LAW

Just about everyone has heard of the Health and Safety at Work etc Act 1974 (HSAW) and quite a few will have heard of the Factories Act 1961, and the Offices, Shops, and Railway Premises Act 1963. The latter two, and indeed virtually all health and safety legislation, are made under the HSAW Act. Rather fewer, however, will have come across the Electricity at Work Regulations 1989. The purpose of this article is not to shock the living daylights out of those of you who enjoy the opportunity to put on display stations or those who provide emergency communications, but to ensure that you know what legally is expected of you, what you should do to comply, and what might happen if you don't and it all goes wrong.

AT HOME

First, however, let's consider your radio installation at home. Within limits, you can do almost what you want, as far as health and safety is concerned.

The HSAW Act does not extend to the individual at home as far as a hobby is concerned. But a word of caution! If your radio installation causes death, injury, or loss to persons on your property – even if they are there without your permission – and it can be shown that your installation is well below the standard expected by the reasonable man on the street, there is the possibility that your installation could be viewed as a 'man trap'. Setting or installing any type of man trap is illegal under the Offences Against the Person Act and other legislation. An example might be, for instance, biasing your HF long wire with 500V to stop anyone attempting to steal it!

Every year, many radio amateurs put on demonstration stations. These may be at fêtes, craft exhibitions, schools, clubs, and so forth. I am sorry to say that of the dozens I have visited over the years, at more than half of them I have seen practices that range from the 'could have been better' right up to the 'I'll order the coffin now' scenario. In general terms, the problems stem from two areas: first, electrical safety: second, antenna system erection, positioning and taking down. In reality, I suppose the problems stem from poor organisation and, although it is not a defence, ignorance.

ELECTRICAL SAFETY

We all know that electricity can kill. But do you know how much current you need and for how long it needs to be flowing? "But it's all right," I hear you say, "there's a fuse in the plug!" Well, unfortunately, the fuse will offer the individual very little, if any, protection from electrocution.

WEEKEND PROJECTS

The fuse is there to protect the wiring between it and the supply, not the appliance or the user. The requirements for a circuit protected by a standard BS1362 fuse (the type fitted in a 13A plug), immaterial of the rating of the fuse, is that the fuse will disconnect the current within a maximum time of 0.4s (regulation 413-02-08 of the 16th Edition of the *IEE Wiring Regulations*). For this to occur, the maximum earth loop impedance must not be greater than 2.3Ω. From Ohm's law, for a 240V supply, this would equate to a current of just over 100A, and if the impedance is considerably lower, say 0.024Ω, as it may be if you are close to the distribution grid transformer, you are talking about currents of 10,000A. This is not a fantasy value, it is the magnitude of current that can flow in fault conditions before the fuse ruptures, and is generally called the *prospective short-circuit current* (PSCC).

Most of us are aware of the residual current device (RCD) – *not* to be confused with the earth leakage circuit breaker (ELCB), which is something entirely different and no longer permitted in domestic installations. The RCD parameter used for 'people protection' is that the device will operate within 30ms – many magnitudes faster than a fuse. It should be noted that the RCD is looking for an imbalance of current between the live and neutral and hence, if you become connected between the live and neutral, you will present a balanced load and the RCD will not operate. Usually, however, there is a fortuitous path to earth and the RCD operates.

POTENTIAL PROBLEMS
The usual electrical problems associated with demonstration stations are:

- Excessively long extension leads.
- Plugs and sockets of an inappropriate type for outdoor use.
- No, or inadequate, mechanical protection for extension leads.
- No RCD at the start of the supply point.
- Cables too small for the size of fuse.
- Use of adapters instead of proper plug boards.
- Taped joints, terminal blocks, or match sticks!
- Access to plugs and sockets by the public.
- Poor or no earths.
- No appropriate testing of installation before use.

"But our station runs off 12V car batteries, so it presents no electric shock risk". Maybe. But your car battery *can* produce several hundred amps and a short circuit can cause an explosion, burns or a fire.

Let's take a quick look at some of the problems listed above:

- Long extension leads of low cross-sectional area. Problem: high loop impedance – the fuse may take a long time (greater than 400ms) to rupture or may not blow at all.
- Cable protection – mechanical protection is needed to prevent damage by vehicles, shoes or animals.

SAFETY, OPERATING PRACTICE AND THE LAW

- Plugs and sockets – should be to BS4343 if used outside.
- RCD – should be at the start of the cable run to protect against all possibilities.
- Adapters – usually provide poor connections, particularly the earth pin. Use a plug board.
- Connections – if you haven't got the correct connector, stop! Taped joints, terminal blocks, and similar 'quick fixes' equate to 'quick death'. The public, and in particular children, must not be able to access plugs, sockets, or switchgear.
- Earthing – this must be adequate for the purpose. A meat skewer of 7/.22 is of no use whatsoever! You need a 1m earth rod and 10mm^2 cable, and the rod must be 900mm in the ground.
- Testing – check that the safety system works. Are L and N the correct way round? Is the ground impedance low enough? Is the installation out of harm's way?

ANTENNA SAFETY

"How on earth can an antenna be unsafe?" Whilst 'on earth' it may be much safer than it is on the top of a 20m tower, but even when it is lying on the ground you must be aware of trip hazards. A crossed Yagi on the floor presents something akin to the defences used in the Second World War to stop tanks. It can easily stop a human being.

Antenna safety covers a number of functions and conditions, including erection, use, and dismantling. Each of these activities brings with it a range of hazards.

A primary operation of any antenna erection is to survey the proposed site well ahead of the day. You should be looking for the proximity of other overhead structures and, in particular, cables. Due regard needs to be given to the route the feeder will take to the station. Does it need to be given protection from vehicles and pedestrians? If open-wire feeders are used they must be well out of reach. Guy wires should be visible (ie marked with hazard-marking tape) or fenced off. Stakes should be similarly marked, eg with traffic cones. Feeder wires and guy wires can easily garotte someone! The antenna system itself needs to be properly attached to the mast. Bailing twine and insulation tape are not appropriate! High-quality clamps in good condition and tightened with proper spanners represent the only satisfactory method.

A car battery can cause an explosion, burns and a fire.

Erecting the mast/tower should be practised well before the day. Written instructions should be available and followed. There should be nobody in the area without a reason to be there. All those involved must wear hard hats and appropriate footwear. One person should be declared in charge and (s)he is the only one who should be giving instructions. You need to consider what will happen if it all goes wrong. Is there plenty of clearance if the mast falls over? Is it immune from the attentions of children? Are warning signs posted? Have you got an adequate first-aid kit immediately to hand? And finally, make sure you have permission to erect the antenna where you intend to put it. During use, a periodic inspection of the antenna and mast should take place. This should alert you to anything which has become loose either on its own or due to unwelcome visitors. It also gives you a chance to ensure that someone else hasn't tied guy ropes, bunting, or floodlights to part of *your* antenna system.

Dismantling the antenna system is usually the reverse process of putting it up. But, here again, many of the problems involved with the erection are present and the same precautions need to be taken.

WHAT IF IT ALL GOES WRONG?
If an accident occurs, the initial reaction is often to rush in and see if you can help. Whilst the intent is laudable, the practice isn't. The first step you should take is to ensure that *you* are not in any danger. It's no use grabbing hold of someone who is in contact with the mains unless you want to join them, or trying to remove someone from under a collapsed mast when the rest of the antenna system is going to fall on top of you. Make the incident site safe. Give whatever first-aid is appropriate. Keep people away unless they have a reason to be there.

Call for help from the emergency services, inform the event organiser and, depending on the nature of the event, get the organiser to contact the Health and Safety Executive (HSE). If you have a camera, or better still a video recorder, record the salient parts of the incident site. Make notes of what happened and don't let anyone interfere with the site or take home 'souvenirs'. If possible get the personal details of the injured party and pass to the ambulance crew and the police. Do not release these details to anyone else. It is the duty of the police to advise relatives of the accident.

LEGAL ACTION
If the HSE and/or police are involved, it is their decision if a prosecution is warranted. As far as the HSE is concerned, it will be looking for breaches of the legislation mentioned at the beginning of this article. The police would generally be interested if someone had caused criminal damage, eg cut through the guy ropes which had led to the accident. This in turn could lead to further charges including manslaughter.

Depending on the seriousness of the offence, it can either be tried at the Magistrates' Court or referred to the Crown Court. For breaches of H & S legislation, the Magistrates' Court can fine up to £25,000. The Crown Court has unlimited powers, including imprisonment. Of course, things may not end there. The injured party (or relatives in the event

SAFETY, OPERATING PRACTICE AND THE LAW

of a death) may decide to pursue a case in the civil court for substantial damages. In all cases, the defence will be to show and prove that all reasonable due care had been taken and diligence shown, that the event had been thought through and planned, and that consideration had been given as to what action to take if it went wrong and appropriate plans made.

It is worth remembering that you cannot legislate for fools or the ignorant. The best tack here is to take out an insurance policy – but then you did that in the initial planning stage, didn't you?

SCREENING – WHAT IS IT AND WHY IS IT IMPORTANT?

Screening or shielding, as it is sometimes known, can be very much a practical exercise when it becomes necessary to restrict a field or fields, close to their source, or alternatively to prevent a field or fields from reaching a sensitive point in a circuit. However, unless the underlying principles are applied, the outcome may not be successful. Textbooks are vague on the subject, so perhaps the following few 'rules of thumb' may go some way toward a better understanding.

As far as this feature is concerned, there are two fields, magnetic and electric. They can and do exist independently but, the instant one of them changes, an electromagnetic field is produced which is able to re-produce itself and propagate into the surrounding space (an electromagnetic wave), or possibly be constrained to travel along a transmission line of some description. For example, if a DC source is switched briefly into a transmission line of any length, the capacitance of the line causes a charging current to flow. This, along with the applied voltage, creates an electromagnetic wave which, willy-nilly, has to set off down the line.

Only *changing* or *alternating* magnetic or electric fields create electromagnetic waves. At low frequencies, eg at 50Hz (mains supply), the radiation is very small, as it is also at audio frequencies. The field surrounding the secondary of a mains transformer of a modern solid-state transceiver would be almost entirely magnetic, whereas the field surrounding the supply terminals of a valve linear power supply (2kV @ <1A) would be almost entirely electric.

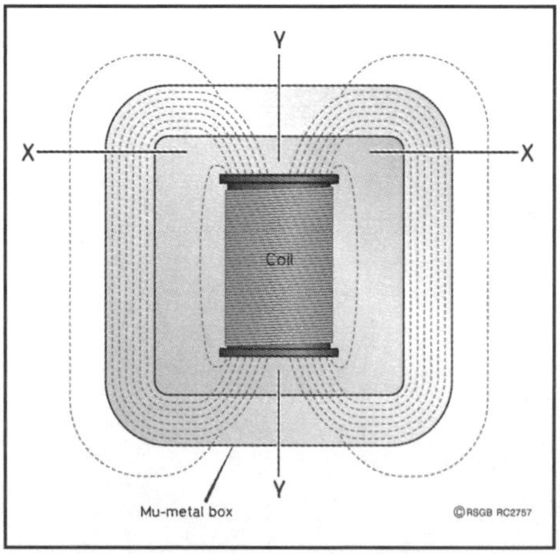

MAGNETIC SCREENING
Now to the question of screening. There are basically two methods available:

(1) diverting the path of the field; and
(2) cancelling it out with an opposing field.

As an example of (1), consider Fig 1. Here the magnetic flux of the coil finds an easy path through the high permeability of the enclosure, leaving only a very small residual field outside. It should be remembered that the lines depicting the path of the magnetic flux do not mean that the flux is made up of lines... it is

Fig 1. The high-permeability box concentrates the flux within the walls. Remember not to join the box along the line X-X.

SCREENING – WHAT IS IT AND WHY IS IT IMPORTANT?

spread throughout the space, though only very very thinly beyond the enclosure.

For the enclosure to be reasonably light in construction, the material used must have a very high permeability. It is here that an alloy called *mu-metal* is employed, though whether this is readily available nowadays is uncertain. At one time microphone transformers were all enclosed in mu-metal boxes, to prevent low and audio frequencies being picked up by the windings. Note that this method *increases* the inductance of the coil. A further very important point is to realise that the join in the box should *not* be made along the line X-X; to do this would insert a high resistance (reluctance) path to the flux. The join should be made along a line Y-Y instead. A mu-metal enclosure is effective from 0Hz to the higher audio frequencies, but as frequency increases the permeability of the alloy diminishes and the resistance increases, so method (2) is employed.

For method (2), copper or aluminium is used. Either material has a low resistance to current flow though the relative permeability is that of air, ie 1, hence it cannot offer a path for the magnetic flux. What it does do is act as a short-circuited secondary to the coil, and the current induced produces a field which tends to cancel out the field produced by the coil. This reduces the inductance of the coil but, provided that the screen is made large enough, the effect can be tolerated and/or allowed for in the design of the coil and screen. The screen diameter should be twice the coil diameter, and the ends of the coil should not

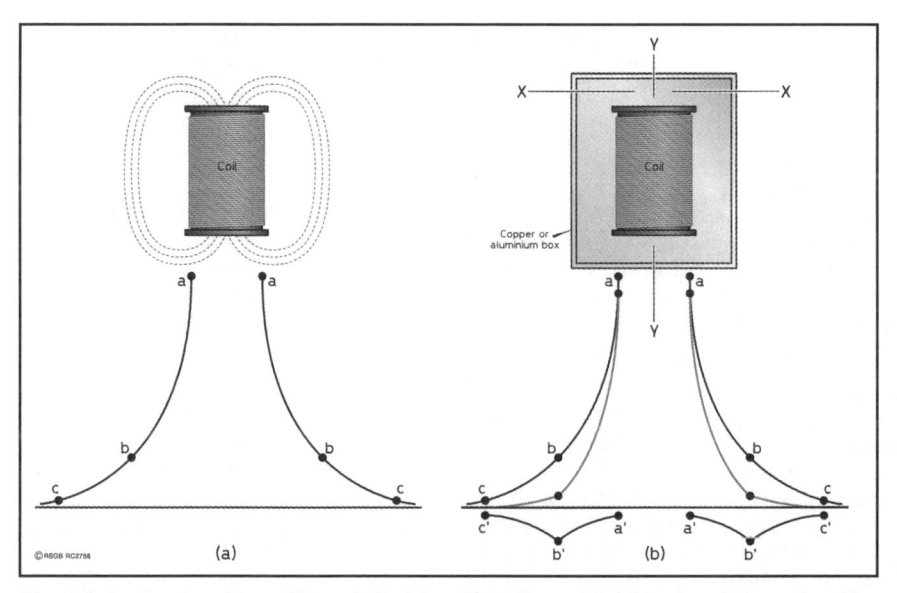

Fig 2(a). A simple coil and field. The flux density is shown by the graph being greatest at 'a' and least at 'c'. (b) The coil is now enclosed in a metal box. Below the line the graph shows the flux set up in the box; the letters 'a', 'b' and 'c' indicating the flux density. Note that it is the reverse of that in the coil. If we add the graphs above and below the horizontal axis, we derive the inner curve above the axis.

WEEKEND PROJECTS

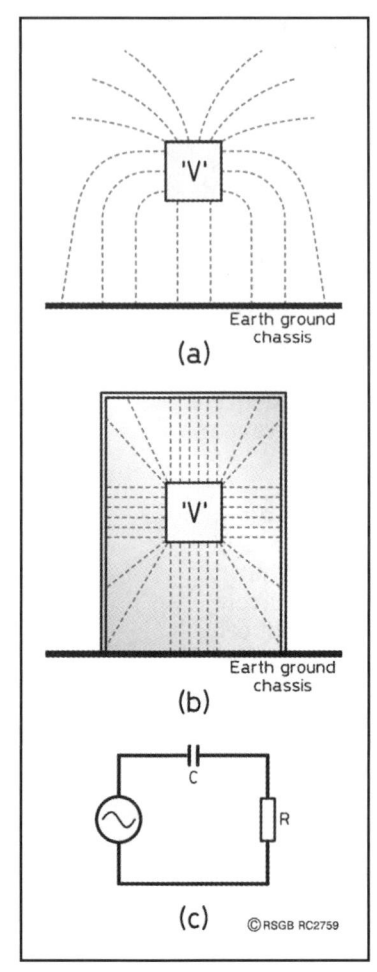

Fig 3(a). The point 'V' is at high potential and surrounded by an electric flux or field. **(b)** The field or flux is contained inside the screen. Every part of the screen is at earth potential, consequently the field or flux is no longer present outside. **(c)** The AC generator represents an HF or VHF source of voltage. Due to the presence of the screen, some capacitive coupling is certain. Should the screen be of poor conductivity, a potential difference will develop across it, giving some loss.

come within one diameter of the ends of the screen. Fig 2(a) shows a coil with the normal flux lines depicting a magnetic field. The letters 'a', 'b', and 'c' indicate the falling-off of flux density as one moves away from the coil, the field being strongest at 'a' and much lower at 'c', though still assumed to be greater than wanted. We assume the current in the coil is alternating, therefore when the screen is placed around the coil, forming a closed circuit, currents will also be produced in it, though their direction will be opposite to those creating them (Lenz's Law). Provided that the screen resistance is very low, the currents produced in it will tend to reduce the flux at 'b' *almost* to zero, the small difference being that which is needed to create the reverse flux, 'b', see Fig 2(b). The letters 'a', 'b' and 'c' below the axis of the graph show the level of the flux present due to the screen at these points – note that they are in the reverse direction to those causing them and very slightly less in amplitude at 'b' and 'c' and very much less at 'a'. Note also that the flux at 'a' will only marginally affect the flux in the coil, therefore the inductance is not greatly reduced. It should be pointed out that the direction of the current in the screen is very different from that of the flux in Fig 1; it is in a continuous ring, parallel to the turns of the coil, hence the joint in this screen may be made in the direction X-X, but not in the direction Y-Y (which would prevent current flowing). As frequency is increased, the depth of field penetration into the screen becomes less, so relatively thin copper or aluminium sheet may be used, but for best results the boxes or enclosures must be watertight (the ideal), requiring joints to be lapped and special care taken with lids, covers, etc. See [1] and [2].

Incidentally, method (2) is completely useless for static or DC fields, and is not successful at low frequencies, therefore use mu-metal for DC and low frequency and copper or aluminium for RF screening.

ELECTRIC SCREENING

Perhaps you will have noticed that nothing has been said about 'earth' so far. This is because earthing has nothing to do with magnetic screening. However, it has everything to do with electric screening. It is the conductivity that is important. To screen a DC or LF electric field all you need to do is to enclose the field in an *earthed* can, box or enclosure.

Fig 3(a) represents a high voltage point, as may be found in a valve linear's power supply. Between this point and chassis/earth there will be an electric field, depicted by the dotted lines. This means that every point in space between 'V' and chassis/earth is at some

SCREENING – WHAT IS IT AND WHY IS IT IMPORTANT?

potential (and hence not zero). If this field is enclosed in a conducting screen connected to chassis/earth, then the screen is considered to be at the potential of the chassis/earth, all points on the screen are at chassis/earth potential, so there is no longer an electric field outside the screen. This is true for non-varying or LF fields, but if the potential at 'V' is varying at a high frequency then there will be a capacitive current between 'V' and the screen, which could cause a potential difference across the screen, so some loss is possible. In practice it is advisable to have the screen as far away as possible from 'V', to avoid too much capacitive loading of the screened high voltage point.

It is not necessary to have a continuous metal screen, it may be made up of individual conductors in the form of a cage.

Although important, the joints need not have the perfection of those in a magnetic screen, so lids etc may be just a push fit. The degree of screening is measured by taking the ratio of the field strength before and after the screen is fitted. Screen effectiveness below 20dB, poor; from 20 to 80dB, average; 80 to 120dB, above average; above 120dB, cost problems.

REFERENCES
[1] *Practical RF Handbook*, by Ian Hickman. Newnes, 1993. Good Appendix.
[2] *Circuit Designer's Companion*, by Tim Williams. Newnes, 1993. Pages 248-252.

SPEECH PROCESSING

There is nothing like a demonstration to illustrate a point, so I would strongly recommend this experiment. On the FM waveband, tune in BBC Radio 4 and adjust the volume to give a comfortable listening level on ordinary speech. Then, retune to a commercial 'pop' station. Whether the content is music or speech, you will probably reach for the control to turn down the volume. The reason for this is the subject of this article, although it is its application to amateur radio that concerns us most. Just like us (but perhaps even more so), the broadcasting companies must not over-modulate or over-deviate, so why is speech on one broadcast so much louder than on another?

FIRST STEPS

The waveform (as might be shown on a cathode ray oscilloscope) of male speech, taken from a discussion programme on BBC Radio 4, is shown in Fig 1(a). Fig 1(b) shows the waveform of male speech from a commercial radio station, and Fig 1(c) the waveform of music taken from the same station. All three recordings were made with the same audio gain setting (with no audio AGC), and are of the same length, so they are thus directly comparable. The equipment used was an Icom IC-PCR1000 receiver, a Soundblaster® PC sound card, and proprietary software for recording wave files. To analyse these waveforms quickly, I wrote a short program in Visual Basic 6.0, which analyses the waveform and calculates the power (in arbitrary units) in each waveform. The results are summarised thus:

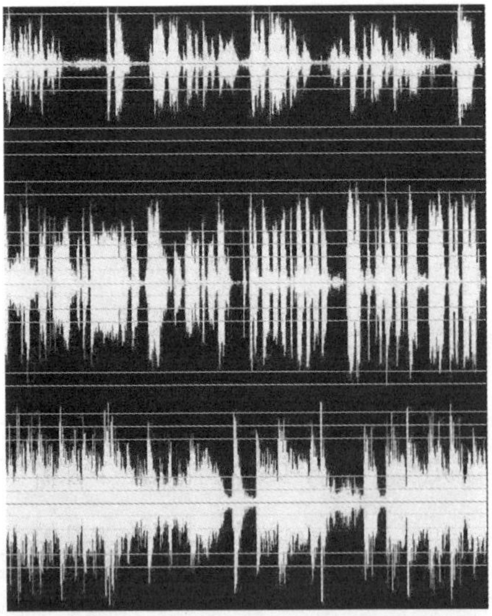

Radio 4 speech: Power = 138
Commercial radio speech: Power = 504
Commercial radio music: Power = 522

For these random selections of programme output, you can see that there is almost four times as much power in commercial radio speech as there is in Radio 4 speech, and that the music played on commercial radio is only marginally louder than the DJ's pearls of wisdom! This process of 'making everything louder' is called *dynamic range compression, contrast compression* or, simply, *compression*. As we amateurs are concerned only with speech, we know it as

Fig 1. (a) A section of male speech from BBC Radio 4. (b) A section of male speech from a commercial radio station. (c) Vocal pop music from the same commercial radio station.

SPEECH PROCESSING

speech compression or *speech processing* and, for reasons to be discussed later, we use it only when working on SSB.

THE PROBLEM
Speech is remarkably 'spiky' in nature, as Figs 1(a) and 1(b) show when viewed on a cathode-ray oscilloscope. This spikiness causes us problems when we have to set the modulation level on our transmitters. On sideband transmissions, over-modulation causes splatter at the very least, and on FM transmissions it produces over-deviation. Both of these must be avoided at all costs.

To do this, we must set the transmitter drive so that the *peaks* of our audio waveform do not over-modulate. To illustrate this, use Fig 1(a), and lay a ruler on it horizontally so that it *just* touches the biggest negative-going peak (we choose a negative-going rather than a positive-going peak, so that the ruler does not hide the waveform). You will now notice one thing – that the average modulating voltage (compared with our self-imposed maximum) is very low. In practical terms, our transmission will lack 'punch', and will not be heard very well under difficult conditions.

SOME SOLUTIONS
In very basic terms, what we need to do is to turn up the audio gain when the speech is quiet and turn it down when the speech is loud. Unfortunately, these variations of loudness occur very quickly, often between syllables, and the technique is something which cannot be done manually.

Two basic methods are used, one of which involves using an amplifier of non-linear transfer characteristic; the other uses automatic control of the audio gain. It should be understood at the outset that, whatever technique is used, it *distorts* the speech waveform. This distortion is (hopefully) controlled so as to improve the intelligibility, even if the result is sometimes unnatural.

Using a non-linear transfer characteristic is, in theory, probably the most attractive solution. A transfer characteristic is the graphical representation of the output voltage from a device compared with the input voltage. Normally this should be a straight line, indicating that the output and input voltages are directly proportional.

To achieve the desired compression, look at the characteristic of Fig 2; notice that, for example, a small input of 0.1V will produce an output of about 0.5V (a gain of about 5), 0.2V will produce an output of about 0.6V (a gain of about 3), whereas a large input of 1V will produce an output of 1V (a gain of unity). Provided the input voltage never exceeds 1V, the circuit will produce an increasing gain at lower voltages. The main problem with a technique such as this is that a circuit to implement it is quite difficult to design. It does have

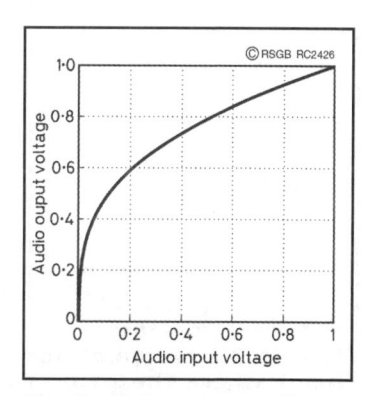

Fig 2. Using a non-linear transfer characteristic as the basis of a speech processor.

WEEKEND PROJECTS

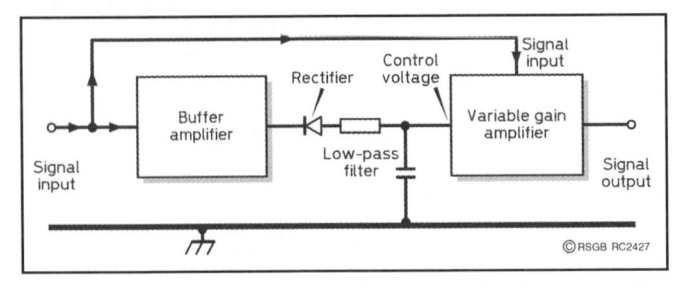

Fig 3. Block diagram of a simple audio AGC circuit.

the advantage (see later) that nothing *physically* changes as a result of changing input levels.

The other techniques are simpler to implement, but have their own disadvantages. Manually altering the audio gain control may be completely impractical, but circuits can be made which will achieve the same result automatically.

The operations involved are straightforward:

(a) Rectify the audio voltage. The average value of an audio signal is zero, so rectification is needed to 'lop off' the lower half of the signal so that it *does* have a mean value.

(b) Feed the rectified signal into a low-pass filter, a simple connection of one resistor and one capacitor (see Fig 3). The output from this filter is the mean value of the signal over short time periods, the audio components having been short-circuited by the capacitor.

(c) This varying mean value is then used to control the gain of another audio amplifier, the gain being reduced as the mean value increases. A block diagram of the circuit is shown in Fig 3. A buffer amplifier is a device which helps to separate the rectifier and filter from preceding circuits and to supply enough gain to operate the rectifier cleanly.

A simple gain control circuit, eminently suitable as the basis for experimental work, is shown in Fig 4. An n-channel JFET is used as one of a pair of feedback resistors in a standard non-inverting op-amp circuit. Connected this way, the resistance between the JFET drain and source depends upon the voltage applied to its gate. The overall gain [1] is adjustable between 1 and 1000 by changing the voltage at the control input.

ATTACK AND DECAY

While Fig 4 is quite a simple and acceptable circuit to use for automatic audio gain control, the derivation of the control voltage to operate it can be quite tricky. The simplest circuit, comprising the rectifier, capacitor and resistor shown in Fig 3, is a good starting point. The values of the resistor and capacitor need to be varied in order that the circuit will act quickly to reduce the gain when a loud sound suddenly appears (this is known as 'attack'), yet will take a little longer to return to its initial gain (known as 'decay') in case another loud sound follows on quickly. Getting the attack and decay right is

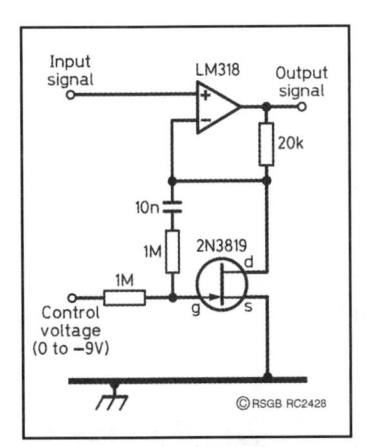

Fig 4. A simple amplifier circuit where the gain is dependent upon the control voltage. The circuit supply voltage should be around ±15V.

SPEECH PROCESSING

quite an art, and a circuit more complex than that of Fig 3 is usually needed.

THE PROBLEMS
No circuit like this is without its problems. Correct attack and decay are obvious candidates, and the solution is somewhat subjective, because no two people speak in exactly the same way. However, the overriding problem with all AGC-based speech processing circuits is that a loud sound *must* get through to the output *before* the circuit can react to reduce the gain. This means that, in unfortunate cases, transient peaks which are louder than they ought to be will get through to the output and could still cause transmitter over-modulation.

The overall drive must be backed off a little to allow for this. Nevertheless, speech processing, when properly used, is of great benefit to the SSB station working in crowded band conditions.

In professional circuits using this principle, the signal between the main input and the input to the gain-controlled amplifier in Fig 3 is subject to a short delay; which means that the gain *can* be turned down *just in time* for the delayed signal which caused it to reach the amplifier!

OTHER METHODS
Speech processing may also be achieved (and more effectively so) by performing the control operation on the RF signal, or by using DSP. Both these techniques are well beyond the scope of this article, however.

FINALLY
We have not discussed speech processing in the context of FM. FM is essentially a short-distance mode, and interference from adjacent stations is not a problem, thus removing the need for processing.

If circuits are useful, you can bet that they have been made into integrated circuits! The speech processor is no exception, and one such device is the VOGAD in an 8-pin DIL chip, the SL6270 (Maplin order code UM73Q).

REFERENCE
[1]   *The Art of Electronics*, Horowitz and Hill, Cambridge University Press, 1988, p240/41.

YOUR FIRST USE OF A REPEATER

This comprehensive article explains why repeaters are necessary for good, consistent contacts on VHF and UHF from moving vehicles, operation which we know by the term 'mobile' or '/M'.

Not many radio articles start with "Once upon a time" but this one does! "Once upon a time, in two far-off valleys, there lived two tribes, separated by a mountain. On this mountain lived a great giant, who forbade any contact between the two peoples unless it was by his permission. For a member of one tribe to speak with a member of the other, they had to call upon the giant to relay their message.

"Now the giant was a sleepy giant, and often did not hear members of the two tribes calling him. To overcome this difficulty, members of each tribe decided to blow upon a whistle to alert the giant, who would then wake up and repeat their messages across the great mountain.

"Some tribesmen, being impatient to wait their turn from the giant (because he would only take one message at a time), tried to make contact directly. However, such was the distance between the two valleys, that only the tribesmen with the loudest voices could make themselves heard, and then only by climbing as high above the valley floor as possible.

"So the people of the two valleys found their giant of such great help, that, as the days went by, other giants were invited to come to the great mountain and pass messages to and fro."

This little tale shows what a repeater does. The giant is the repeater, and the tribesmen are individual amateurs, mobile, driving around. And yes, repeaters are often located, if not on a mountain, then certainly on high ground.

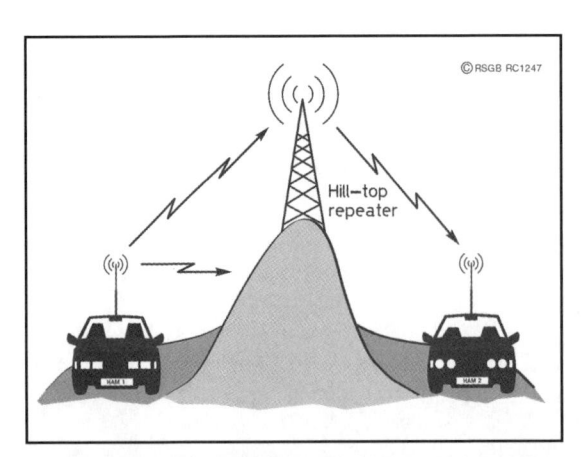

Fig 1. Because of the terrain, the two mobiles would not be able to achieve communication from one to the other without the assistance of the hill-top repeater.

WHAT IS A REPEATER?

A repeater is a device which will receive a signal on one frequency, and simultaneously transmit it on another frequency. Careful design has meant that repeaters can receive and transmit

YOUR FIRST USE OF A REPEATER

within the same band. This means that the same aerial can be used for both reception and transmission. In effect, the receiving and transmitting coverage of the mobile station becomes that of the repeater and, since the repeater is favourably sited on high ground or a tall mast (Fig 1), the range is greatly improved over that of unassisted, or 'simplex', operation (Fig 2). The coverage areas of two mobile stations is always changing shape, whereas the coverage of a repeater will stay constant, and can even be published.

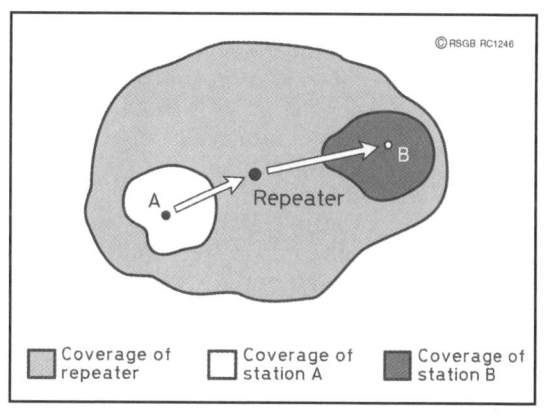

Fig 2. The simplex coverage areas of mobile stations A and B are constantly changing as the two vehicles pass through different terrain. The repeater allows communication between A and B anywhere within the repeater's coverage area.

HISTORICALLY SPEAKING

It is in the early 1970s that the history of amateur repeaters begins in the UK. The influx of compact, Japanese-origin transceivers operating on narrow-band FM allowed amateurs for the first time to fit such units easily in their cars. Previously, more cumbersome valve-operated sets had been used, but were hardly conducive to easy communications.

With some experience gained from the private mobile radio (PMR) business, it was soon realised how a strategically-sited repeater could allow relatively low-powered sets in cars to communicate over relatively long distances. The first repeater in England was GB3PI, located near Cambridge, which was soon followed by others on both 2m and 70cm.

Realising that there was potential for a nationwide network of repeaters, the RSGB created the forerunner of the Repeater Management Group (RMG) to oversee the specifications, planning, and vetting of applications.

The actual building of the repeaters was undertaken by local groups of amateurs, and a whole new 'club' infrastructure was created of repeater groups recruiting members for an annual subscription. The groups would build, maintain and operate repeaters in their areas on behalf of their members.

Although use of the repeaters was open to all licensed amateurs, those who used the repeaters would be morally expected to join their local group, or to subscribe in some way. Some groups were very well organised, publishing professional newsletters, holding regular meetings, and making many technical developments on their repeater systems.

At the core of each repeater group was the technical team who would have access to the necessary equipment to keep these repeaters operating within the required specifications.

WEEKEND PROJECTS

TECHNICAL CONSTRAINTS

Technically, a repeater consists of a receiver, a transmitter, aerials, filtering, and control logic (Fig 3). Each repeater is allocated a 'channel', consisting of a receive frequency and a transmit frequency within the same band, but separated by 600kHz (on 2m) or 1.6MHz (on 70cm). A 1750Hz toneburst on the input frequency was employed to 'wake up' the giant – sorry, repeater – when it would then relay the input audio to the transmitter. The identifying callsign (of the format GB3xx) of the repeater is transmitted in Morse, and if no audio is detected after, say, 10 seconds, the transmitter carrier shuts down.

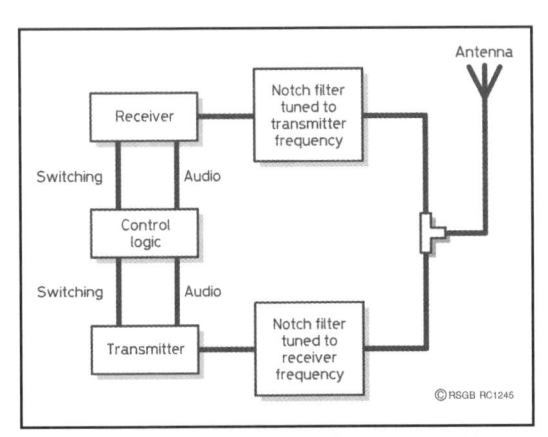

Fig 3. Block diagram of the 70cm repeater GB3BD in Bedfordshire: the receiver is at the bottom, logic circuitry is in the middle and transmitter is at the top.

The major technical constraint to the repeater builders over the years has been the relatively narrow spacing in frequency between transmit and receive. You need a very sensitive receiver working (often from the same aerial) with a transmitter only 600kHz away. To overcome this problem, considerable filtering is required, and repeaters will almost invariably employ cavity notch filters to reduce this desensitization, so the repeater does not gain the unenviable reputation of being 'deaf'.

The siting of a repeater is important, and often the major constraint for repeater builders after the filtering. Nowadays, prime radio sites are in great demand, and site owners can charge commercial rates for space. Fortunately, amateur repeaters were around long before cellular phones, and started before the explosion in mobile radio services in the last 10 to 20 years. This has meant that many amateurs have managed to secure favourable status with a number of mast owners. Besides, many licensed amateurs and repeater builders are themselves working in the industry.

Despite all that, getting permission to use a good radio site can prove difficult, and more than one repeater has gone under because of site problems. There are repeaters on many bands, from proposals to use 29MHz right up to 10GHz. There are also several specialist repeater projects. There is a pilot single-sideband (PSSB) repeater near Buxton in Derbyshire, and there are several FM television repeaters on the 1296MHz band. There are also linked repeaters and repeaters linked to the Internet, and CTCSS tone access is now gathering favour (but that is another subject!)

There has, regrettably, been a down side to repeaters as well. Being sited, as they have to be, on prime hilltop locations, they have sometimes attracted unauthorised use, and some repeaters, predominantly in urban areas, have suffered abuse over the years.

YOUR FIRST USE OF A REPEATER

USING REPEATERS

Using repeaters involves *duplex* operation: you transmit and receive on different frequencies. Most available transceivers have this facility built in. There is a user's protocol. Firstly, repeaters are primarily intended for mobile or portable users. It is definitely unacceptable to 'hog' them for long lengths of time. Always welcome newcomers to join in, encourage membership of the group, and remember your normal amateur protocols are as valid on repeaters as on simplex operation (use of callsign, courtesy etc). Before attempting to transmit, ensure that:

- Your transmitter and receiver are on the correct frequencies (remember the repeater split).
- Your tone access (if fitted) is operating correctly.
- Your peak deviation is set correctly (some repeaters will not relay your signal if it is incorrect). Any adjustments you have to make should be done into a dummy load, *not* on-air.

Avoid using the repeater from your base station; it is really intended for the benefit of local mobile and portable stations. If you really do intend to try it from a fixed station, use the lowest power to get into the repeater (under 1W is usual in the majority of situations where you can hear the repeater well). Always listen before transmitting. Unless you are calling another specific station, simply announce that you are 'listening through', eg "GM8LBC listening through GB3CS". On the other hand, if you are responding to someone specific, try something like, "GM0ZZZ from GM8LBC". Once contact is established:

- At the beginning and the end of each over, you need give only your own callsign, eg "from GM8LBC".
- Change frequency to a simplex frequency at the first opportunity, especially if you are operating from a fixed station.
- Keep your overs short and to the point, or they may time-out, and do not forget to wait for the 'K' or 'blip' if the repeater uses one.
- Do not monopolise the repeater when busy as others may be waiting to use it.
- If your signal is very noisy into the repeater, or if you are only opening the repeater squelch intermittently, finish the contact and try again later.

JOINING UP

Repeaters have initiated many newcomers to the amateur radio hobby. Repeater outputs can be monitored easily, either with widely available amateur equipment, or perhaps with a scanner. Repeater groups themselves have been able to join up prospective users, and through local meetings, or contacts, newcomers have been shown how to proceed. Many amateurs use repeaters in addition to their other amateur activities, perhaps using their local repeaters whilst travelling to and from work, then doing something entirely different in the evenings.

WEEKEND PROJECTS

Amateur 'purists' have, particularly in the early years of repeater growth, shunned their existence, as not being truly in the 'ham spirit'. 30 years on, this is very much a minority view. In my personal case, I started with the hobby just as repeaters were starting. I received my licence in the mid-1970s, just as the networks were in the early stages. I had my first QSOs on FM, and have been a member of various repeater groups over the years, worked though dozens, helped to build and maintain several, and now help to administer them on behalf of the RSGB.

It is the strength of our hobby that has presented so many opportunities for devotees to find their niches. Repeater builders are amongst the most technically competent and experienced in the hobby; many are employed in the PMR industry or in professional communications. Repeaters are often co-sited with major broadcasters or PMR users, so have to be of a high technical standard.

MORE INFORMATION
There is not a lot of additional reading about repeaters, but lists of the current UK networks are available in the annual *RSGB Yearbook*. A letter to the Chairman, RMG, c/o RSGB HQ will be directed to an appropriate member of the Repeater Management Group for reply.

It is also recommended that you should find out about your local repeater group and join it, thereby opening up many other possibilities for learning more about amateur repeaters. Your local repeater group may publish information, perhaps in the form of newsletters or information sheets.

The RSGB *Amateur Radio Operating Manual* has a whole chapter devoted to mobile and portable operation, including operation through repeaters.

INDEX

10GHz WBFM transceiver	36
80m transceiver	47
1750Hz toneburst	151

A

Attenuator, switched	64
ATU	
L-match	112
T-match	148
Audio level indicator	108

B

Baluns	170
Bargraph display	108
Battery alarm	70
Battery tester	72
Bulletin boards	130

C

Capacitance meter	76
Cathode-ray oscilloscope	179
Cathode-ray tube	175
Charger for NiCad batteries	114
Computer / radio interfaces	101
Fully isolated	101
Simple	103
For handheld radios	104
Corner reflector aerial for UHF	25
Counterpoise	12
CTCSS	151

D

Diode / Transistor tester	84
Dipole	
Centre-fed	15
Coaxial feeder	16
Multi-band	11
Open-wire feeder	16
Single-band	10
Diodes for protection	182
Dip oscillator	86
Dual-voltage supply	90
DXpedition	185

E

Earth-continuity tester	80
Field-strength indicator	97

F

Filter, audio	67
Frame aerial for HF	99

G

Gunn oscillator	34, 35
Frequency pushing	36

H

HF contesting guide	193

I

Injector, signal	137
IOTA, beginners' guide	204

J

J-pole for 50MHz	171

K

Keyer, electronic	94

L

Long-wire	12
Loop alarm	110
Loop, magnetic	20

WEEKEND PROJECTS

M

Maintenance, aerial	14
Match	
Gamma	2, 4

N

Noise reduction	210

O

Op-amp tester	118
Optical communication	123

P

Packet radio principles	128
Photometer and polar diagram	213
'Piano keyer', touch-sensitive	105
Plumber's Delight	4
Potential divider, loading effects	90
Power supply, portable	132

Q

QSLs, how to manage your	216
Quad for 6m	2

R

Radiation resistance	221
Repeaters	242
RF mixing	208
RF probe, amplified	135
RTTY contests, a beginners' guide	225

S

Safety, operating practice the law	229
Satellites	
Converter	40
Telemetry	44
Up- and down-links	39
Screening	234
Magnetic	234
Electric	236
Shoestring, getting started on a	189
S-meter for DC receivers	140
Speech processing	238
Standing-wave indicator (HF)	165

T

Terminal node controller	128
Thyristor	110
Time-out unit	145
Transmission lines (feeders)	15

V

Voltage follower	91
Voltage monitor	155
Voltage regulation	159

W

Wattmeter	162
Wheatstone bridge	80

Y

Yagi	
6m	4
Tube, portable, for 144MHz	28

Z

Zener diode	159